"十三五"普通高等教育本科系列教材

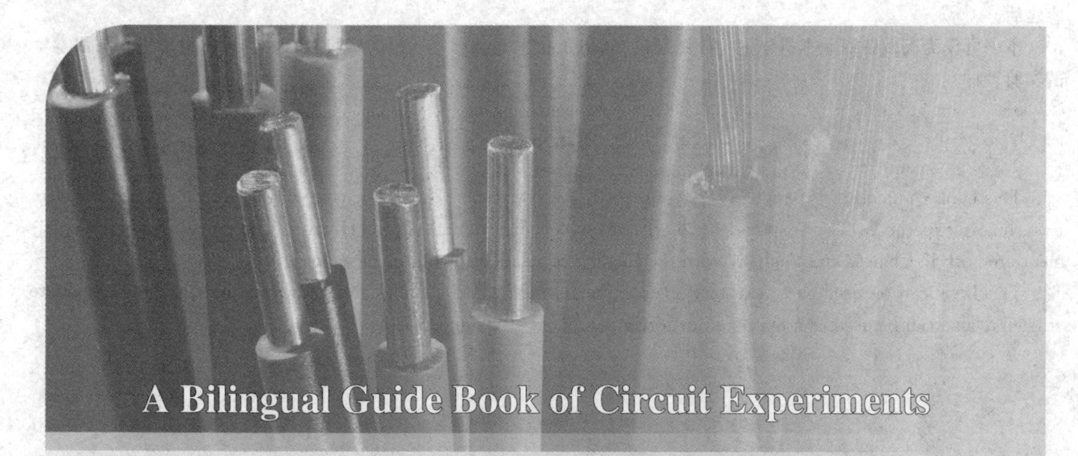

A Bilingual Guide Book of Circuit Experiments

电路实验双语指导书

主　编　刘　骁
副主编　陈　艳
编　写　陈攀峰　王民富
主　审　胡　钋

中国电力出版社
CHINA ELECTRIC POWER PRESS

内 容 提 要

本书为“十三五”普通高等教育本科系列教材。全书共包含26个实验项目，其中实验1～实验22为电路电工基础内容，实验23～实验26为继电接触控制内容。每个实验内容均为中英文双语对照，英文内容放在中文内容之后。

本书可作为高校各专业本科学生“电路理论”“电工技术基础”等课程的实验教材，也可作为来华外国留学生的实验教材。

Synopsis

This book contains 26 experiments, experiment 1 ～ experiment 22 are basic contents of electric circuits and electrotechnics, experiment 23 ～ experiment 26 are contents of relay contact control circuits. The contents of each experiment are both in Chinese and English, with the English contents after the Chinese ones.

This book can be used as a textbook of experiments in Electric Circuits and Electrotechnics for undergraduate students, also can be used as a textbook for foreign students studying in China.

图书在版编目（CIP）数据

电路实验双语指导书/刘骁主编．—北京：中国电力出版社，2019.2（2023.7重印）
“十三五”普通高等教育本科规划教材
ISBN 978-7-5198-2807-3

Ⅰ.①电… Ⅱ.①刘… Ⅲ.①电路—实验—双语教学—高等学校—教材 Ⅳ.①TM13—33

中国版本图书馆CIP数据核字（2018）第295329号

出版发行：中国电力出版社
地　　址：北京市东城区北京站西街19号（邮政编码100005）
网　　址：http://www.cepp.sgcc.com.cn
责任编辑：牛梦洁（mengjie-niu@sgcc.com.cn）
责任校对：黄　蓓　太兴华
装帧设计：赵丽媛
责任印制：钱兴根

印　　刷：北京雁林吉兆印刷有限公司
版　　次：2019年2月第一版
印　　次：2023年7月北京第八次印刷
开　　本：787毫米×1092毫米　16开本
印　　张：13
字　　数：306千字
定　　价：38.00元

前　言

双语教学是我国现阶段高等教育改革的趋势之一。目前电路理论课程已有学校编写了双语教材，电路实验课程作为电路理论的辅助课程，尚未发现通行的双语教材，故编者编写该双语实验教材。

本书作为一般本科生电路、电工学实验双语教材的同时，也可方便学校开展留学生教学。编者长期在华北电力大学电气与电子工程学院从事电工实验室工作，为本校多个专业的本科生以及来自多个国家的留学生讲授独立设课的电路实验课程。在对外国留学生授课的过程中，编者认识到对于掌握汉语较困难的留学生，一部双语指导书可对其理解实验内容有所帮助。

本书中部分实验单独设置，但实际教学过程中学生可以在标准的 2 学时实验中一并完成。为保证实验课程内容充实，本书对这部分实验进行了合并处理，特此说明。本书基本内容由刘骁编写，由陈艳审定。陈攀峰、王民富对部分内容进行了修改和补充。武汉大学胡钋教授在审稿后，给予了大量宝贵的意见。编者在此表示衷心感谢。

由于编者水平所限，书内中英文内容如有不妥之处，望读者批评指正。

编　者

2018 年 11 月于华北电力大学

Preface

Bilingual teaching is one of the trends for reforms in Chinese higher education. Bilingual electric circuits textbooks have been already compiled by some schools, electric circuits experiments is the auxiliary course of electric circuits and there are no bilingual textbooks for it have been published yet, so the author wishes to compile this book.

This book can be used as the textbook of experiments in Electric Circuits and Electrotechnics for undergraduate students, at the same time it also can be a convenience in the foreign students' education. The author of this book works in Electrical Laboratory, School of Electrical & Electronic Engineering, North China Electric Power University for a long term, teaches the independent course of electric circuits experiments for native students in multiple majors and foreign students from multiple countries. During the course of teaching foreign students, the author realized that for the foreign students who have difficulties in mastering Chinese, a bilingual guide book will provide some help in understanding the contents of the experiments.

Some experiments in this book are set as independent lessons traditionally, but in actual teaching process, they can be finished in 2 class hours, in this book these experiments are consolidated to ensure the enrichment of experiment lessons' contents. The basic content of this book is compiled and translated by Liu Xiao, and examined ang approved by Chen Yan . Chen Panfeng and Wang Minfu revised and supplemented some content. Prof. Hu Po gives plenty of valuable suggestions after the reviewing, the author wishes to express the sincere gratitude here.

The author's capability is limited. It is hoped that the readers will kindly point the errors.

The Author

2018.11 North China Electric Power University

目 录

实验 1　电路元件伏安特性的测量

一、实验目的

（1）掌握线性电阻、非线性电阻元件伏安特性的逐点测试法。

（2）学习恒压源、直流电压表及电流表的使用方法。

二、实验原理

任一二端电阻元件的特性可用该元件上的端电压 U 与通过该元件的电流 I 之间的函数关系 $U=f(I)$ 来表示，即用 $U-I$ 平面上的一条曲线来表征，这条曲线称为该电阻元件的伏安特性曲线。根据伏安特性的不同，电阻元件分线性电阻和非线性电阻两大类。

线性电阻元件的伏安特性曲线是一条通过坐标原点的直线，如图 1-1(a) 所示，该直线的斜率只由电阻元件的电阻值 R 决定，其阻值为常数，与元件两端的电压 U 和通过该元件的电流 I 无关。非线性电阻元件的阻值 R 不是常数，即在不同的电压作用下，电阻值是不同的。常见的非线性电阻如白炽灯泡、二极管、稳压二极管等，它们的伏安特性如图 1-1 中(b)～(d)所示。在图中，$U>0$ 的部分为正向特性，$U<0$ 的部分为反向特性。

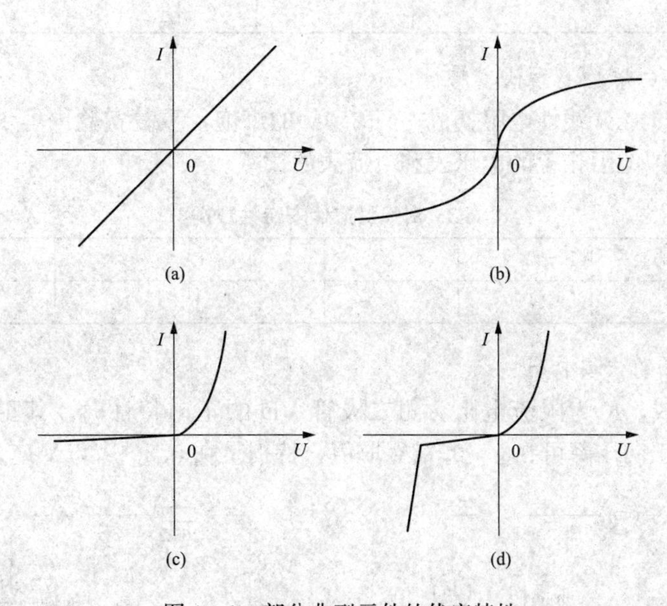

图 1-1　部分典型元件的伏安特性

(a) 线性电阻；(b) 白炽灯丝；(c) 普通二极管；(d) 稳压二极管

绘制伏安特性曲线通常采用逐点测试法，即在不同的端电压作用下，测量出相应的电流，然后逐点绘制出伏安特性曲线，根据伏安特性曲线便可计算其电阻值。

三、实验设备

实验设备见表 1-1。

表 1-1　　实　验　设　备

设备名称	型号与规格	数量	实验模块[1]
恒压源	0～30V	1	NDG-02
直流电压表	0～200V	1	NDG-03
直流电流表	0～2000mA	1	
电阻	1kΩ	1	NDG-13
电阻	200Ω	1	
白炽灯泡	6.3V	1	
二极管	IN4007	1	
稳压二极管	IN4728	1	

四、实验内容

1. 测定线性电阻的伏安特性

按图 1-2 接线，调节恒压源的输出电压U，从 0V 开始缓慢地增加，不能超过 10V，在表 1-2 中记下相应的电压表和电流表的读数。

表 1-2　　线性电阻伏安特性数据

U (V)	0	2	4	6	8	10
I (mA)						

2. 测定 6.3V 白炽灯泡的伏安特性

将图 1-2 中的 1kΩ 线性电阻换成一只 6.3V 的灯泡，重复实验内容 1.，电压不能超过 6.3V，在表 1-3 中记下相应的电压表和电流表的读数。

表 1-3　　6.3V 白炽灯泡伏安特性数据

U (V)	0	1	2	3	4	5	6
I (mA)							

3. 测定二极管的伏安特性

按图 1-3 接线，R 为限流电阻。测二极管 VD 的正向特性时，其正向电流不得超过 25mA，二极管的正向压降可在 0～0.75V 取值，特别是在 0.5～0.75V 应多取几个测量点。

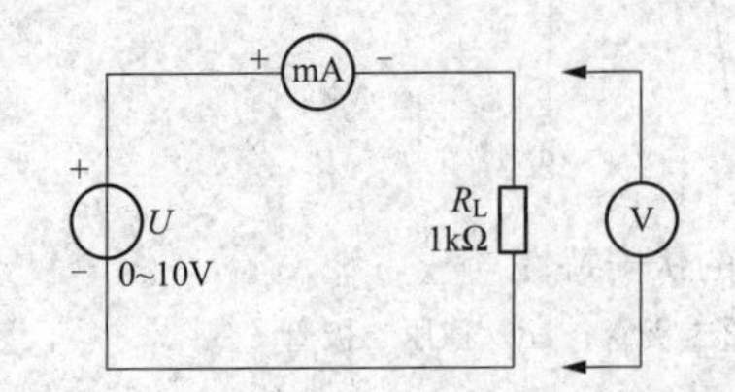

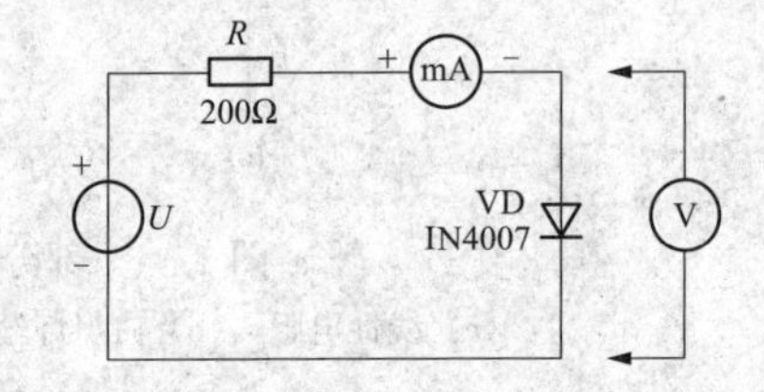

图 1-2　测定线性电阻的伏安特性　　图 1-3　测定半导体二极管的伏安特性

测反向特性时，将恒压源的输出端正、负连线互换，调节恒压源输出电压U，反向电压

[1] 实验模块型号来自于编者所在实验室使用的实验装置。

不能超过−30V，将数据分别记入表 1－4 和表 1－5 中。

表 1－4　　二极管正向特性实验数据

U（V）	0	0.2	0.4	0.45	0.5	0.55	0.60	0.65	0.70	0.75
I（mA）										

表 1－5　　二极管反向特性实验数据

U（V）	0	−5	−10	−15	−20	−25	−30
I（mA）							

4. 测定稳压管的伏安特性

将图 1－3 中的二极管 IN4007 换成稳压管 IN4728，重复实验内容 3. 的测量，其正、反向电流不得超过±20mA，将数据分别记入表 1－6 和表 1－7 中。

表 1－6　　稳压管正向特性实验数据

U（V）	0	0.2	0.4	0.45	0.5	0.55	0.60	0.65	0.70	0.75
I（mA）										

表 1－7　　稳压管反向特性实验数据

U（V）	0	−1	−1.5	−2	−2.5	−2.8	−3	−3.2	−3.5	−3.55
I（mA）										

五、注意事项

（1）恒压源输出端切勿短路。

（2）测量前应事先估算电压和电流值，合理选择仪表量程，勿使仪表超量程。

六、思考题

（1）线性电阻与非线性电阻的伏安特性有何区别？它们的电阻值与通过的电流有无关系？

（2）如何计算线性电阻与非线性电阻的电阻值？

（3）在图 1－3 中，设 $U=2V$，$U_{VD+}=0.7V$，则毫安表读数为多少？

（4）设某电阻元件的伏安特性函数式为 $I=f(U)$，在绘制伏安特性曲线时应如何在坐标系内取点？

七、实验报告

（1）根据实验数据，分别在方格纸上绘制出各个电阻的伏安特性曲线。

（2）根据伏安特性曲线，计算线性电阻的电阻值，并与实际电阻值比较。

（3）根据伏安特性曲线，计算白炽灯在额定电压(6.3V)时的电阻值，当电压降低 20％时，阻值为多少？

Experiment 1 Measurement of Volt-Ampere Characteristics of Circuit Elements

● **Objectives**

1. Learn the point-by-point testing method for volt-ampere characteristics of linear and nonlinear resistors.

2. Learn how to use constant voltage source, DC voltmeter and ammeter.

● **Principles**

The characteristic of any resistance element can be represented by the function $U = f(I)$, and U and I are the voltage and the current of the resistor respectively. Also the characteristic can be expressed as a curve on the $U - I$ plane, and this curve is called volt-ampere characteristic curve. The resistance elements are divided into two categories: linear resistor and nonlinear resistor in terms of volt-ampere characteristic.

The volt-ampere characteristic curve of linear resistance element is a straight line passing through the origin of plane, as shown in Figure 1 - 1. The slope of this straight line is determined only by the resistance value R of resistor. This resistance value is a constant which is irrelevant to the voltage U and the current I of the element. The resistance value R of nonlinear resistance element is not a constant, which means the value is different at different voltages. The volt-ampere characteristics curves of common nonlinear resistance elements such as incandescent lamp, diode, zener diode, as shown in Figure 1 - 1(b), (c) and (d). In the Figure, the part of $U > 0$ is forward characteristic and the part of $U < 0$ is reverse.

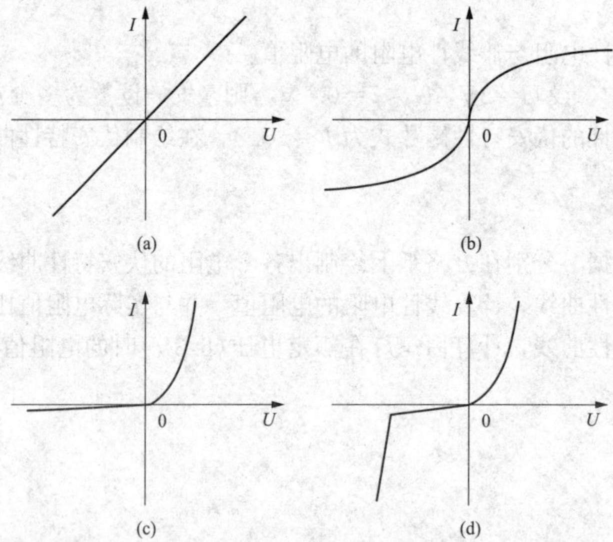

Figure 1 - 1 The Volt-Ampere Characteristies of Some Typical Circuit Elements

(a) Linear Resister; (b) Incandescent Filament; (c) Common Diode; (d) Zener Diode

The point-by-point testing method is used where a volt-ampere characteristic curve is drawn. Step to get volt-ampere curve: ①Measure the current at different terminal voltages. ②Draw the volt-ampere characteristic curve point-by point. The resistance value can be calculated from the curve.

- **Equipment**

Equipment is shown in Table 1 - 1.

Table 1 - 1 Equipment

Equipment	Model or Specification	Quantity	Module❶
Constant Voltage Source	0~30V	1	NDG - 02
DC Voltmeter	0~200V	1	NDG - 03
DC Ammeter	0~2000mA	1	
Resistor	1kΩ	1	NDG - 13
Resistor	200Ω	1	
Incandescent Lamp	6. 3V	1	
Diode	IN4007	1	
Zener Diode	IN4728	1	

- **Contents**

1. Measure the Volt-Ampere Characteristic of a Linear Resistor

Connect the circuit according to Figure 1 - 2, and slowly adjust the output voltage U from 0V to max 10V. Fill in Table 1 - 2 with the data from the voltmeter and ammeter.

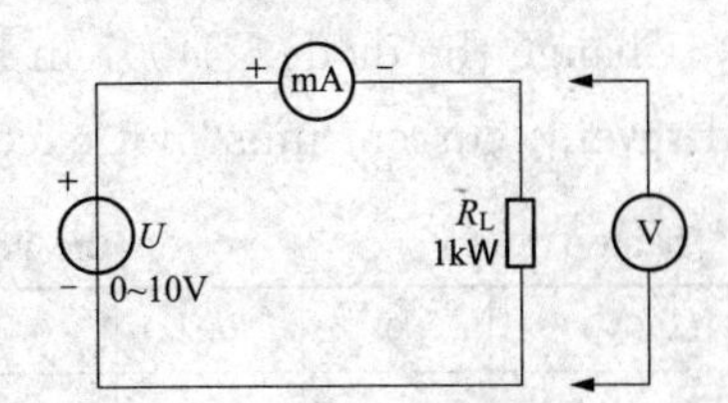

Figure 1 - 2 Measure the Volt -Ampere Characteristic of a Linear Resistor

Table 1 - 2 the Volt-Ampere Characteristic of Linear Resistor

U (V)	0	2	4	6	8	10
I (mA)						

2. Measure the Volt-Ampere Characteristic of 6. 3V Incandescent Lamp

Change the 1kΩ linear resistor in the Figure 1 - 2 to a 6. 3V light bulb, repeat step 1, the voltage must not exceed 6. 3V. Fill in Table 1 - 3 with the data from the voltmeter and ammeter.

Table 1 - 3 the Volt-Ampere Characteristic of 6. 3V Incandescent Lamp

U (V)	0	1	2	3	4	5	6
I (mA)							

3. Measure the Volt-Ampere Characteristics of Diode

Connect the circuit according to Figure 1 - 3, where R is a current-limiting resistor. The

❶ The types of the modules are from the experiment modules used in the author's lab.

forward current must not exceed 25mA when the forward characteristic of diode VD is measured. The forward voltage value of diode should be in the 0～0.75V interval, and more measuring point should be set in the 0.5V～0.75V interval. Change the positive and negative terminals of constant voltage source when the reverse characteristic is being measured. Adjust the output of voltage source, the reverse voltage must not exceed −30V. Fill in Table 1-4 and Table 1-5.

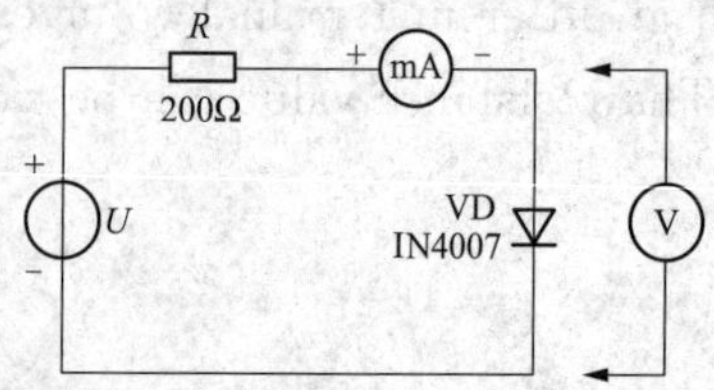

Figure 1-3 Measure the Volt-Ampere Characteristics of Diode

Table 1-4 **the Forward Characteristic of Diode**

U (V)	0	0.2	0.4	0.45	0.5	0.55	0.60	0.65	0.70	0.75
I (mA)										

Table 1-5 **the Reverse Characteristic of Diode**

U (V)	0	−5	−10	−15	−20	−25	−30
I (mA)							

4. Measure the Volt-Ampere Characteristics of Zener Diode

Change the diode IN4007 in Figure 1-3 to zener diode IN4728, repeat step 3. The forward and reverse current must not exceed ±20mA. Fill in Table 1-6 and Table 1-7.

Table 1-6 **the Forward Characteristic of Zener Diode**

U (V)	0	0.2	0.4	0.45	0.5	0.55	0.60	0.65	0.70	0.75
I (mA)										

Table 1-7 **the Reverse Characteristic of Zener Diode**

U (V)	0	−1	−1.5	−2	−2.5	−2.8	−3	−3.2	−3.5	−3.55
I (mA)										

● **Notes**

1. The output terminals of constant voltage source must not be short-circuited.

2. Roughly estimate the values of voltage and current before measuring them, and choose the range in terms of the estimated values.

● **Questions**

1. What is the difference between the characteristics of linear and nonlinear resistors? Do the resistance values of them related to the current of these elements?

2. How to calculate the resistance values of linear and nonlinear resistors?

3. In Figure 1-3, letting $U = 2\text{V}$, $U_{D+} = 0.7\text{V}$, what is the reading of mA ammeter?

4. Letting the function of volt-ampere characteristic of some resistor be $I = f(U)$, how to take points in the coordinate plane?

● **Experiment Report**

1. Draw the volt-ampere characteristic curves of resistors on graph paper.

2. Calculate the resistance value of the linear resistor according to the volt-ampere characteristic curve and compare it with the actual value.

3. Calculate the resistance value of incandescent lamp at rated voltage (6. 3V) according to the volt-ampere characteristic curve. What is the resistance value when the voltage is reduced by 20%?

实验2 电位、电压的测量，基尔霍夫定律和叠加、齐次原理

一、实验目的

（1）学会测量电路中各点电位和电压的方法，理解电位的相对性和电压的绝对性。

（2）学会电路电位图的测量、绘制方法。

（3）验证基尔霍夫定律，加深对基尔霍夫定律的理解。

（4）验证线性电路的叠加原理，加深对线性电路的叠加性和齐次性的认识和理解。

（5）掌握恒压源、直流电压表及电流表的使用方法，学会使用电流插头、插座测量各支路电流的方法。

（6）掌握检查、分析电路简单故障的能力。

二、实验原理

（1）在一个确定的闭合电路中，各点电位的大小视所选的电位参考点的不同而异，但任意两点之间的电压（即两点之间的电位差）则是不变的，这一性质称为电位的相对性和电压的绝对性。据此性质，可用一只电压表来测量出电路中各点的电位及任意两点间的电压。

若以电路中的电位值作纵坐标，电路中各点位置作横坐标，将测量到的各点电位在该坐标平面中标出，并把标出点按顺序用直线相连接，就可得到电路的电位图。电位图中每一段直线段即表示该两点电位的变化；任意两点的电位变化，即为该两点之间的电压。

在电路中，电位参考点可任意选定，对于不同的参考点，所绘出的电位图形是不同的，但其各点电位变化的规律是一样的。

（2）基尔霍夫电流定律（KCL）和电压定律（KVL）是电路的基本定律，它们分别描述节点电流和回路电压。对电路中的任一节点而言，在设定电流的参考方向下，应有 $\Sigma I=0$，一般流出节点的电流取负号，流入节点的电流取正号。对任何一个闭合回路而言，在设定电压的参考方向下，绕行一周，应有 $\Sigma U=0$，一般电压方向与绕行方向一致的电压取正号，电压方向与绕行方向相反的电压取负号。

运用上述定律时必须注意各支路或闭合回路中电流的正方向，此方向可预先任意设定。

（3）叠加原理指出：在有多个独立源共同作用下的线性电路中，通过每一个元件的电流或其两端的电压，可以看成是由每一个独立源单独作用时在该元件上所产生的电流或电压的代数和。

线性电路的齐次性是指当电路的激励增加或减小 K 倍时，电路的响应也将增加或减小 K 倍。

叠加性和齐次性均只适用于求解线性电路中的电流、电压。对于非线性电路，叠加性和齐次性都不适用。

三、实验设备

实验设备见表 2-1。

表 2-1　实验设备

设备名称	型号与规格	数量	实验模块
恒压源	0～30V	1	NDG-02
直流电压表	0～200V	1	NDG-03
直流电流表	0～2000mA	1	
实验电路	基尔霍夫定律和叠加原理	1	NDG-12

四、实验内容

1. 电位、电压的测量

（1）实验步骤。实验电路如图 2-1 所示，分别将两路恒压源接入 U_{S1}、U_{S2}，并将输出电压调到 $U_{S1}=6V$，$U_{S2}=12V$。

以图 2-1 中 A 点作为电位参考点，分别测量 U_{S1+}、B、C、D、U_{S2+} 各点的电位 ϕ，数据记入表 2-2 中；然后以 D 点作为电位参考点，重复上述步骤。

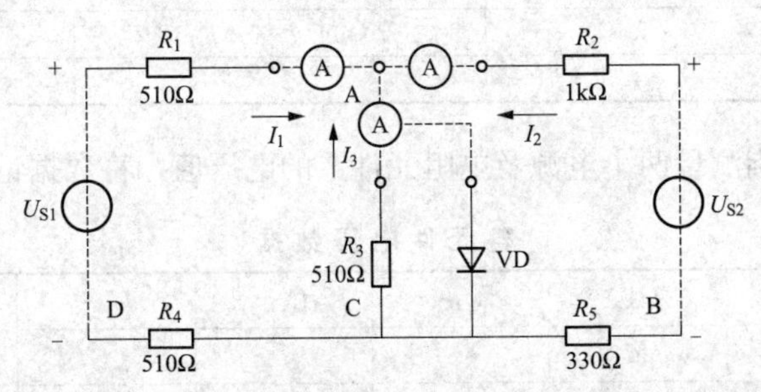

图 2-1　电位、电压的测量及基尔霍夫定律、叠加与齐次性原理验证电路

表 2-2　电路中各点电位数据　(V)

电位参考点	ϕ_A	ϕ_{S2+}	ϕ_B	ϕ_C	ϕ_D	ϕ_{S1+}
A	0					
D				0		

在以 A 点和 D 点为参考点时，分别测量电压 U_{AS2+}、U_{BC}、U_{CA}、U_{CD}、U_{S1+A}，测量数据记入表 2-3 中。

表 2-3　电路中各点电压数据　(V)

电位参考点	U	U_{AS2+}	U_{S2}	U_{BC}	U_{CA}	U_{CD}	U_{S1}	U_{S1+A}
A	理论值		12				6	
	测量值							
	相对误差							
D	理论值		12				6	
	测量值							
	相对误差							

（2）注意事项。

1）使用直流电压表测量电位时，用－端接入参考电位点，＋端接入被测各点。

2）使用直流电压表测量电压时，将电压表的+端接入被测电压参考方向的+端，一端接入被测电压参考方向的一端，如测量U_{BC}，将电压表+端接入B、一端接入C，反接则为测量U_{CB}。

3）不同参考点下的电位和电压均需单独测量，不能用已测数据代替。

4）计算电压时应按照电路图列式计算，不能用已测电位相减计算电压。

5）恒压源输出电压以电压表读数为准，以下同。

6）实验过程中，不要抓握安装在仪表屏上方的日光灯管，以防其破裂。

2. 基尔霍夫定律的验证

（1）实验步骤。实验电路如图2-1，保持$U_{S1}=6V$，$U_{S2}=12V$。三条支路的电流参考方向如图中的I_1、I_2、I_3所示。

将电流表分别接入三条支路，测量各电流值，记入表2-4中。

表2-4 支路电流数据 （mA）

支路电流	I_1	I_2	I_3
理论值			
测量值			
相对误差			

用直流电压表分别测量两个电源及电阻元件上的电压值，将数据记入表2-5中。

表2-5 各元件电压数据 （V）

U	U_{AS2+}	U_{S2}	U_{BC}	U_{CD}	U_{AC}	U_{S1}	U_{S1+A}
理论值		12				6	
测量值							
相对误差							

（2）注意事项。

1）注意电流参考方向、电流表引线，电流表读数正负之间的对应关系。

2）注意电压降方向，电压表引线，电压表读数正负之间的对应关系。

（3）线性电路的叠加性、齐次性的验证。

1）实验步骤实验电路如图2-1，U_{S1}、U_{S2}调整为$U_{S1}=12V$，$U_{S2}=6V$。按表2-6要求，测量电路中的电流和电压，记入表中。在电源U_{S1}单独作用时，移去U_{S2}，并将该处短接，U_{S2}单独作用及U_{S2}调整到2倍单独作用时U_{S1}同样处理。

表2-6 线性电路实验数据

测量项目 / 实验内容	U_{S1} (V)	U_{S2} (V)	I_1 (mA)	I_2 (mA)	I_3 (mA)	U_{AS2+} (V)	U_{BC} V)	U_{AC} (V)	U_{CD} (V)	U_{S1+A} (V)
U_{S1}单独作用	12	0								
U_{S2}单独作用	0	6								
U_{S1}，U_{S2}共同作用	12	6								
2倍U_{S2}单独作用	0	12								

将电阻R_3切换为二极管IN4007，重复上述步骤，数据记入表2-7中。

表 2-7　非线性电路实验数据

实验内容 \ 测量项目	U_{S1} (V)	U_{S2} (V)	I_1 (mA)	I_2 (mA)	I_3 (mA)	U_{AS2+} (V)	U_{BC} V)	U_{AC} (V)	U_{CD} (V)	U_{S1+A} (V)
U_{S1}单独作用	12	0								
U_{S2}单独作用	0	6								
U_{S1}，U_{S2}共同作用	12	6								
2倍U_{S2}单独作用	0	12								

2）注意事项。

a. 防止恒压源短路。

b. 线性电路实验内容不能接入二极管。

五、思考题

（1）在用数字电压表、电流表测量电位、电压、电流时，为何数据前会出现负号，这表示什么意义？

（2）在图 2-1 的电路中可以列几个 KVL 方程？它们与绕行方向有无关系？

（3）若用指针万用表直流毫安挡测各支路电流，什么情况下可能出现指针反偏，应如何处理，在记录数据时应注意什么？

（4）在表 2-6 中，各电阻消耗的功率能否用叠加原理计算得出？

六、实验报告

（1）根据测量到的电位数据，分别绘制出参考点为 A 点和 D 点的两个电位图形。

（2）计算表 2-3～表 2-5 中理论值及相对误差。

（3）根据表 2-4 中实验数据，选定任一节点，验证基尔霍夫电流定律（KCL）的正确性。

（4）根据表 2-5 中实验数据，选定任一回路，验证基尔霍夫电压定律（KVL）的正确性。

（5）根据表 2-6 中实验数据，验证线性电路的叠加性和齐次性。

（6）分析表 2-7 中实验数据，说明叠加性和齐次性是否适用该实验电路。

Experiment 2 Measurement of Potential and Voltage, Kirchhoff's Laws, Additivity and Homogeneity of Linear Circuits

● **Objectives**

1. Learn the method of measuring the electric potential and voltages in an electric circuit, understand the relativity of potential and the absoluteness of voltages in electric circuits.

2. Learn how to measure potential in an electric circuit and draw the potential diagram.

3. Validate and understand Kirchhoff's Laws.

4. Validate and understand the additivity and homogeneity of linear circuits.

5. Learn how to use constant power source, DC voltmeter and DC ammeter, grasp the method of measuring branch current with current plugs and sockets.

6. Be able to exam and analyze the simple faults in electric circuits.

● **Principles**

1. In a certain closed circuit, the potential of each node are different due to the changes of reference node, but the voltage (the potential difference) across any two nodes doesn't change with reference node. Those properties called the relativity of potential and the absoluteness of voltage. Therefore we can measure the potential of each node and voltage across any two nodes with a voltmeter.

If take the potential values as ordinate and the nodes in circuits as abscissa, mark the potential value of each node in this coordinate plane, and connect the marked nodes with straight lines in order, the potential diagram is drawn. Every straight line in the potential diagram indicates the change of potential between the nodes at both ends of the line. The difference of potential between any two nodes is the voltage between them.

The reference node in a circuit can be chosen arbitrarily, the potential diagram is different according to different reference nodes, but the change rules of potential are the same.

2. Kirchhoff's Current Law (KCL) and Kirchhoff's Voltage Law (KVL) are basic laws of electric circuits. They describe the node current and loop voltages respectively. For any node in a circuit, there should be $\Sigma I=0$ with the reference directions of all the current. In general, the current flowing out of a node is negative and the current flowing into a node is positive. For a closed loop, there should be $\Sigma U=0$ around the loop with the reference directions of all the voltages. In general, the voltages consistent with the direction of the loop are positive and the voltages opposite to the direction of the loop are negative.

The positive direction of the current in the branches or closed loops must be noticed when the laws above are used. The direction can be assigned arbitrarily.

3. The superposition principle states that: the voltage across (or current through) and

element in a linear circuit is the algebraic sum of the voltages across (or current through) that element due to each independent source acting alone.

The homogeneity property for a linear circuit states that if the input (also called the excitation) is multiplied by a constant, the output (also called the response) is multiplied by the same constant.

The superposition and homogeneity are only suitable for solving current and voltages in linear circuits.

● Equipment

Equipment is shown in Table 2 - 1.

Table 2 - 1 Equipment

Equipment	Model or Specification	Quantity	Module
Constant Voltage Source	0～30V	1	NDG - 02
DC Voltmeter	0～200V	1	NDG - 03
DC Ammeter	0～2000mA	1	
Experiment Circuit	Kirchhoff's Laws And Superposition Principle	1	NDG - 12

● Contents

1. Potentials and Voltages

(1) Contents

The experiment circuit is shown in Figure 2 - 1, connect the two constant voltage source U_{S1} and U_{S2} to the circuit and adjust the output voltages to $U_{S1} = 6V$ and $U_{S2} = 12V$.

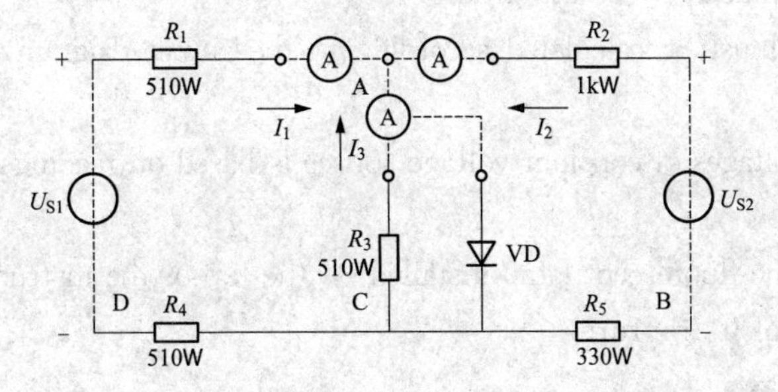

Figure 2 - 1 The Circuit of Measurement of Potential and Voltage, Kirchhoff's Laws, and the Verification of Additivity and Homogeneity

With the node A as the potential reference node, measure the potential ϕ of the points U_{S1+}, B, C, D, U_{S2+}, fill in Table 2 - 2, change the potential reference node to the point D, repeat the previous steps.

Table 2 - 2 Potential of Each Node of the Circuit (V)

Potential Reference Node	ϕ_A	ϕ_{S2+}	ϕ_B	ϕ_C	ϕ_D	ϕ_{S1+}
A	0					
D				0		

Measure the voltages U_{AS2+}, U_{BC}, U_{CA}, U_{CD}, U_{S1+A} when the point A and D are set as the potential reference node, fill in Table 2 - 3.

Table 2 - 3 **Voltages of the Circuit** (V)

Potential Reference Node	U	U_{AS2+}	U_{S2}	U_{BC}	U_{CA}	U_{CD}	U_{S1}	U_{S1+A}
A	Theoretical Value		12				6	
	Measured Value							
	Relative Error							
D	Theoretical Value		12				6	
	Measured Value							
	Relative Error							

(2) Notes

1) When the DC voltmeter is used for measuring the potential, the terminal (−) of the voltmeter should be connected to the potential reference node, the terminal (+) should be connected to the measured nodes.

2) When the DC voltmeter is used for measuring the voltages, the terminal (+) of the voltmeter should be connected to the terminal (+) of the measured voltage, the terminal (−) should be connected to the terminal (−) of measured voltage. For example, when U_{BC} is measured, connect the terminal (+) to the point B, connect terminal (−) to the point C, if the terminals are reversed, the voltage is U_{CB}.

3) The potential and voltages under different potential reference nodes must be measured separately, rather than replaced by existing data.

4) The voltages must be calculated according to the circuit diagram, rather than potential substraction.

5) The output voltages of constant voltage source is based on readings of voltmeter, and the same below.

6) Do not grab the fluorescent lamp installed on the top of the instrument Panel during the experiment, beware of its rapture.

2. Kirchhoff's Laws

(1) Contents

The experiment circuit is shown in Figure 2 - 1, and $U_{S1}=6V$ and $U_{S2}=12V$. The reference directions of the current I_1, I_2, and I_3 are shown in Figure 2 - 1.

Connect the ammeters to the three branches to measure the current, and fill in Table 2 - 4.

Table 2 - 4 **Current of Branches** (mA)

Current	I_1	I_2	I_3
Theoretical Value			
Measured Value			
Relative Error			

Measure the voltages across the two power supplies and resistors, fill in Table 2 - 5.

Table 2 - 5 **Voltages of Elements** (V)

U	U_{AS2+}	U_{S2}	U_{BC}	U_{CD}	U_{AC}	U_{S1}	U_{S1+A}
Theoretical Value		12				6	
Measured Value							
Relative Error							

(2) Notes

1) Pay attention to the relationship among the current reference direction, the ammeter terminals and the positive and negative readings of ammeter.

2) Pay attention to the relationship among the direction of voltage drop, the voltmeter terminals and the positive and negative readings of voltmeter.

3. Verification of the Superposition and Homogeneity of a Linear Circuit

(1) Contents

The experiment circuit is shown in Figure 2 - 1, and adjust U_{S1} and U_{S2} to $U_{S1} = 12V$ and $U_{S2} = 6V$. Measure the currents and voltages in Table 2 - 6 and fill in it. When U_{S1} acts alone, remove U_{S2}, and replace it with a wire, do the same thing to U_{S1} when U_{S2} and U_{S2} which increases by 2 times alone.

Table 2 - 6 **Data of the Linear Circuit**

Data / Content	U_{S1} (V)	U_{S2} (V)	I_1 (mA)	I_2 (mA)	I_3 (mA)	U_{AS2+} (V)	U_{BC} V)	U_{AC} (V)	U_{CD} (V)	U_{S1+A} (V)
U_{S1} Works Alone	12	0								
U_{S2} Works Alone	0	6								
U_{S1}, U_{S2} Work Together	12	6								
Doubled U_{S2} Works Alone	0	12								

Change the resistor R_3 for diode IN4007, repeat the previous steps, and fill in Table 2 - 7.

Table 2 - 7 **Data of the Nonlinear Circuit**

Data / Content	U_{S1} (V)	U_{S2} (V)	I_1 (mA)	I_2 (mA)	I_3 (mA)	U_{AS2+} (V)	U_{BC} V)	U_{AC} (V)	U_{CD} (V)	U_{S1+A} (V)
U_{S1} Works Alone	12	0								
U_{S2} Works Alone	0	6								
U_{S1}, U_{S2} Work Together	12	6								
Doubled U_{S2} Works Alone	0	12								

(2) Notes

1) Avoid short circuit of the constant voltage source.

2) Diode must not be connected to the circuit when the linear circuit experiment is performed.

● **Notes**

1. Why does the symbol "—" appear when the potential, voltage, current are measured? What does it mean?

2. How many KVL equations can be set up in the circuit of Figure 2-1? Are these equations related to the direction of loop?

3. Under what circumstances will the pointer reverse when the current is measured by a multimeter in DC milliampere range? How to deal with this situation? What should attention be paid to when the data is filled in Table?

4. Can the power by the resistors in Table 2-6 be calculated by the superposition principle?

● **Experiment Report**

1. Draw the potential diagrams in terms of the potential data when the points A and D are chosen as reference nodes.

2. Calculate the theoretical values and relative errors in Table 2-3～Table 2-5.

3. Use the data in Table 2-4, choose a node, and verify KCL.

4. Use the data in Table 2-5, choose a loop, and verify KVL.

5. Use the data in Table 2-6, and verify the superposition and homogeneity of linear circuits.

6. Analyze the data in Table 2-7, and explain whether the superposition and homogeneity are applicable to this circuit.

实验3　受控源的研究

一、实验目的

（1）加深对受控源的理解。

（2）了解由运算放大器组成受控源电路的原理，了解运算放大器的应用。

（3）掌握受控源特性的测量方法。

二、实验原理

1. 受控源

受控源向外电路提供的电压或电流受其他支路的电压或电流控制，因而受控源是双口元件：一个为控制端口，或称输入端口，输入控制量（电压或电流）；另一个为受控端口或称输出端口，向外电路提供电压或电流。受控端口的电压或电流，受控制端口的电压或电流的控制。根据控制变量与受控变量的不同组合，受控源分为四类。

（1）电压控制电压源（VCVS），如图3－1(a) 所示，其特性为

$$u_2 = \mu u_1$$

$$\mu = \frac{u_2}{u_1}$$

式中：μ 为转移电压比（即电压放大倍数）。

（2）电压控制电流源（VCCS），如图3－1(b) 所示，其特性为

$$i_2 = g u_1$$

$$g = \frac{i_2}{u_1}$$

式中：g 为转移电导。

（3）电流控制电压源（CCVS），如图3－1(c) 所示，其特性为

$$u_2 = r i_1$$

$$r = \frac{u_2}{i_1}$$

式中：r 为转移电阻。

（4）电流控制电流源（CCCS），如图3－1(d) 所示，其特性为

$$i_2 = \beta i_1$$

$$\beta = \frac{i_2}{i_1}$$

式中：β 为转移电流比（即电流放大倍数）。

2. 用运算放大器组成的受控源

运算放大器的电路符号如图3－2所示，具有同相输入端 u_+ 和反相输入端 u_- 两个输入端，一个输出端 u_o，放大倍数为 A，则 $u_o = A(u_+ - u_-)$。

对于理想运算放大器，放大倍数 A 为∞，输入电阻为∞，输出电阻为0，由此可得出两个特性。

特性 1：$u_{+}=u_{-}$；

特性 2：$i_{+}=i_{-}=0$。

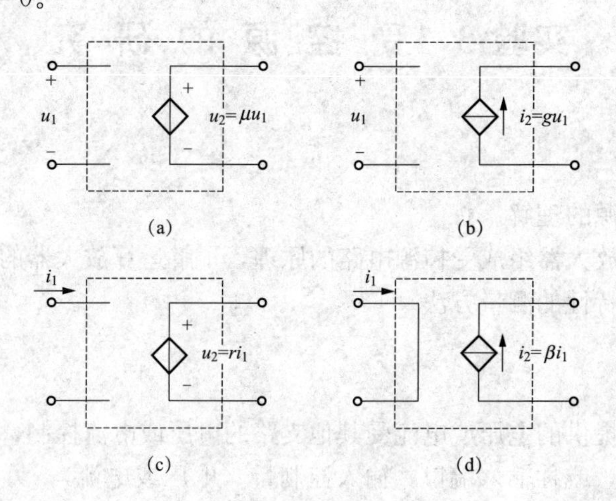

图 3-1　受控源

(a) 电压控制电压源；(b) 电压控制电流源；(c) 电流控制电压源；(d) 电流控制电流源

(1) 电压控制电压源 (VCVS)。电压控制电压源电路如图 3-3 所示。

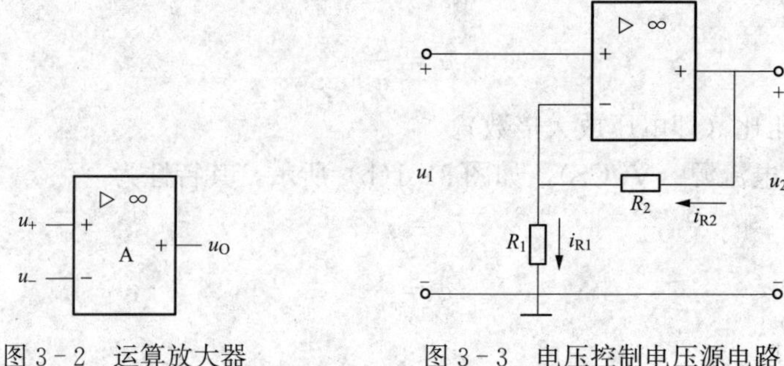

图 3-2　运算放大器　　　　图 3-3　电压控制电压源电路

由运算放大器的特性 1 可知 $u_{+}=u_{-}=u_1$，则

$$i_{R1}=\frac{u_1}{R_1}\qquad i_{R2}=\frac{u_2-u_1}{R_2}$$

由运算放大器的特性 2 可知 $i_{R1}=i_{R2}$，代入 i_{R1}、i_{R2}，得

$$u_2=\left(1+\frac{R_2}{R_1}\right)u_1$$

可见，运算放大器的输出电压 u_2 受输入电压 u_1 控制，其电路模型如图 3-1(a) 所示，转移电压比为 $\mu=\left(1+\frac{R_2}{R_1}\right)$。

(2) 电压控制电流源 (VCCS)。电压控制电流源电路如图 3-4 所示。

由运算放大器的特性 1 可知 $u_{+}=u_{-}=u_1$，则

$$i_R=\frac{u_1}{R_1}$$

由运算放大器的特性 2 可知 $i_2=i_R=\frac{U_1}{R_1}$，即 i_2 只受输入电压 u_1 控制，与负载 R_L 无关

(实际上要求R_L为有限值)。其电路模型如图3-1(b)所示，转移电导为$g=\frac{i_2}{u_1}=\frac{1}{R_1}$。

(3) 电流控制电压源(CCVS)。电流控制电压源电路如图3-5所示。

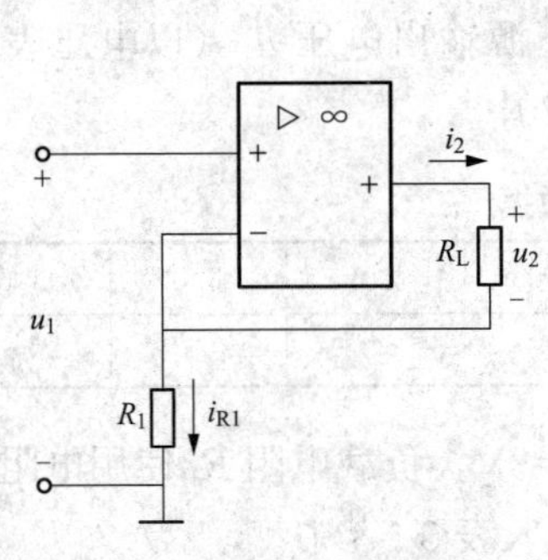

图3-4 电压控制电流源电路

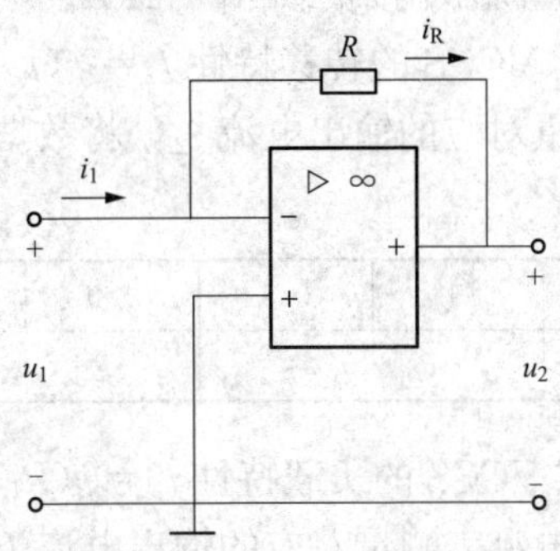

图3-5 电流控制电压源电路

由运算放大器的特性1可知$u_-=u_+=0$，$u_2=Ri_R$①

由运算放大器的特性2可知$i_R=i_1$ ②

将式②代入式①，得

$$u_2=Ri_1$$

即输出电压u_2受输入电流i_1的控制。其电路模型如图3-1(c)所示，转移电阻为$r=\frac{u_2}{i_1}=R$。

(4) 电流控制电流源(CCCS)。电流控制电流源电路如图3-6所示。由运算放大器的特性1可知$u_-=u_+=0$，$i_{R1}=\frac{R_2}{R_1+R_2}i_2$ ③

由运算放大器的特性2可知$i_{R1}=-i_1$ 将式④代入式③，得$i_2=-\left(1+\frac{R_1}{R_2}\right)i_1$④

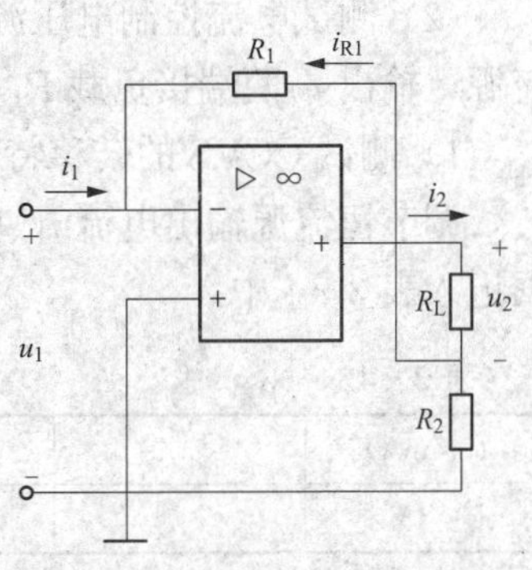

图3-6 电流控制电流源电路

即输出电流i_2只受输入电流i_1的控制。与负载R_L无关。它的电路模型如图3-1(d)所示，转移电流比$\beta=\frac{i_2}{i_1}=-\left(1+\frac{R_1}{R_2}\right)$。

三、实验设备

实验设备见表3-1。

表3-1 实验设备

设备名称	型号与规格	数量	实验模块
恒压源	0~30V	1	NDG-02
恒流源	0~500mA	1	
直流电压表	0~200V	1	NDG-03
直流电流表	0~2000mA	1	
实验电路	受控源		NDG-11
电阻	1kΩ	1	NDG-13
可调电阻箱	0~9999Ω	2	NDG-06

四、实验内容

(1) 测试电压控制电流源（VCCS）特性实验电路如图 3-1(b) 所示，图中 u_1 使用恒压源的输出端，i_2 两端接负载 $R_L=1k\Omega$。

1) 测试 VCCS 的转移特性 $i_2=f(u_1)$。调节恒压源输出电压 u_1（以电压表读数为准），用电流表测量对应的输出电流 i_2，将数据记入表 3-2 中。

表 3-2 VCCS 的转移特性数据

u_1 (V)	0	0.5	1.0	1.5	2.0	2.5	3.0	3.5	4.0
i_2 (mA)									

2) 测试 VCCS 的负载特性 $i_2=f(R_L)$。保持 $u_1=3V$，负载电阻 R_L 使用电阻箱，调节电阻大小，用电流表测量对应的输出电流 I_2，将数据记入表 3-3 中。

表 3-3 VCCS 的负载特性数据

R_L (kΩ)	1	2	3	4	5	6	7	8	9
i_2 (mA)									

(2) 测试电流控制电压源（CCVS）特性。实验电路如图 3-1(c) 所示，图中 i_1 使用恒流源，输出 u_2 两端接负载 $R_L=1k\Omega$。

1) 测试 CCVS 的转移特性 $u_2=f(i_1)$。

调节恒流源输出电流 i_1（以电流表读数为准），用电压表测量对应的输出电压 u_2，将数据记入表 3-4 中。

表 3-4 CCVS 的转移特性数据

I_1 (mA)	0.1	0.15	0.2	0.25	0.3	0.4
U_2 (V)						

2) 测试 CCVS 的负载特性 $u_2=f(R_L)$。保持 $i_1=0.3mA$，负载电阻 R_L 使用电阻箱，调节电阻大小，用电压表测量对应的输出电压 u_2，将数据记入表 3-5 中。

表 3-5 CCVS 的负载特性数据

R_L (kΩ)	1	2	3	4	5	6	7	8	9
u_2 (V)									

(3) 测试电压控制电压源（VCVS）特性。电压控制电压源（VCVS）可由电压控制电流源（VCCS）和电流控制电压源（CCVS）串联而成。实验电路由图 3-1(b)、(c) 构成，将图 3-1(b) 的输入端 u_1 接恒压源的输出端，输出端 i_2 与图 3-1 (c) 的输入端 i_1 相连，图 3-1(c) 的输出端 u_2 接负载 $R_L=1k\Omega$。

1) 测试 VCVS 的转移特性 $u_2=f(u_1)$。调节恒压源输出电压 u_1（以电压表读数为准），用电压表测量对应的输出电压 u_2，将数据记入表 3-6 中。

表 3-6 VCVS 的转移特性数据

u_1 (V)	0	0.5	1.0	1.5	2.0
u_2 (V)					

2）测试VCVS的负载特性$u_2=f(R_L)$。保持$u_1=1V$，负载电阻R_L使用电阻箱，调节电阻大小，用电压表测量对应的输出电压u_2，将数据记入表3-7中。

表3-7 **VCVS的负载特性数据**

R_L（kΩ）	1	2	3	4	5	6	7	8	9
u_2（V）									

（4）测试电流控制电流源（CCCS）特性。电流控制电流源（CCCS）可由电流控制电压源（CCVS）和电压控制电流源（VCCS）串联而成。实验电路由图3-1(c)、(b)构成，将图3-1(c)的输入端i_1接恒流源，输出端u_2与图3-1(b)的输入端u_1相连，图3-1(b)的输出端i_2接负载$R_L=1k\Omega$。

1）测试CCCS的转移特性$i_2=f(i_1)$。调节恒流源输出电流i_1（以电流表读数为准），用电流表测量对应的输出电流i_2并将数据记入表3-8中。

表3-8 **CCCS的转移特性数据**

i_1（mA）	0	0.05	0.10	0.15	0.20	0.25	0.30
i_2（mA）							

2）测试CCCS的负载特性$i_2=f(R_L)$。保持$i_1=0.3mA$，负载电阻R_L使用电阻箱，调节电阻大小，用电流表测量对应的输出电流i_2，将数据记入表3-9中。

表3-9 **CCCS的负载特性数据**

R_L（kΩ）	1	2	3	4	5	6	7	8	9
i_2（mA）									

五、注意事项

（1）运算放大器输出端不能与地短路，输入端电压不宜过高（小于5V）。

（2）用恒流源供电的实验中，不要使恒流源的负载开路。

六、思考题

（1）若受控源控制量的极性反向，试问其输出极性是否发生变化？

（2）受控源的控制特性是否适合于交流信号？

七、实验报告

（1）根据实验数据，在方格纸上分别绘出四种受控源的转移特性和负载特性曲线，并求出相应的转移参量μ、g、r和β。

（2）参考实验数据，说明转移参量μ、g、r和β受电路中哪些参数的影响？如何改变它们的大小？

Experiment 3 Study of Controlled Sources

- **Objectives**

1. Acquire a better understanding of the controlled sources.

2. Know the principle of controlled sources comprised of operational amplifiers, and know the applications of operational amplifiers.

3. Learn how to measure characteristics of controlled sources.

- **Principles**

1. Controlled Sources

Since the control of a controlled source is achieved by a voltage or current of some other element in the circuit, the controlled source is a two-port element: one port is the control port (or the input port) for the control variable (voltage or current); the other port is the controlled port (or the output port) as a voltage or current source. Since the source can be voltage or current, it follows that there are four possible types of controlled sources, namely.

(1) A voltage-controlled voltage source (VCVS), it is shown in Figure 3-1(a), and the characteristic is

$$u_2 = \mu u_1$$

where μ is transfer voltage ratio (or voltage amplification factor).

(2) A voltage-controlled current source (VCCS), it is shown in Figure 3-1(b), and the characteristic is

$$i_2 = g u_1$$

where g is transconductance.

(3) A current-controlled voltage source (CCVS), it is shown in Figure 3-1(c), and the characteristic is

$$u_2 = r i_1$$

where r is transresistance.

(4) A current-controlled current source (CCCS), it is shown in Figure 3-1(d), and the characteristic is

$$i_2 = \beta i_1$$

where β is transfer current ratio (or current amplification factor).

2. Controlled Sources Comprised of Ideal Operational Amplifiers

The circuit symbol for the operational amplifier is shown in Figure 3-2. The operational amplifier has two inputs and one output. The inputs are marked with (−) and (+) to specify inverting and non-inverting inputs respectively. An input applied to the non-inverting terminal will appear with the same polarity at the output, while an input applied to the inverting terminal will appear inverted at the output. The operational amplifier senses the difference between the two inputs, multiples it by the gain A, and causes the resulting voltage to appear at the output. Thus, the output u_o is given by $u_o = A(u_+ - u_-)$.

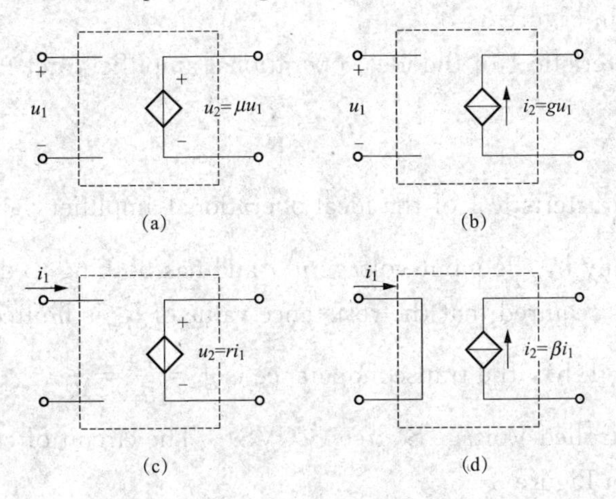

Figure 3 - 1 Controlled Sources

(a) VCVS; (b) VCCS; (c) CCVS; (d) CCCS

For an ideal operational amplifier, the gain A is ∞, the input resistance is ∞, and the output resistance is 0. It follows that two important characteristics of the ideal operational amplifier are

$$u_+ = u_-$$

$$i_+ = i_- = 0$$

(1) The circuit of voltage controlled voltage source is shown in Figure 3 - 3.

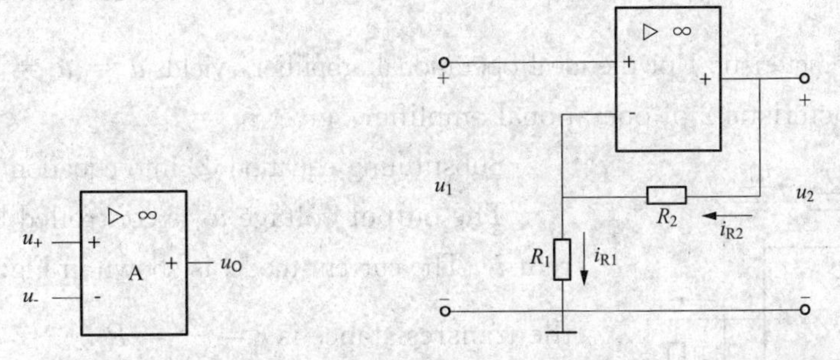

Figure 3 - 2 Operational Amplifer Figure 3 - 3 The Circuit of VCVS

Applying characteristic 1 of the ideal operational amplifier, gives $u_+ = u_- = u_1$, therefore

$$i_{R1} = \frac{u_1}{R_1}, \quad i_{R2} = \frac{u_2 - u_1}{R_2}$$

Using characteristic 2 of the ideal operational amplifier, yields $i_{R1} = i_{R2}$, substitute i_{R1} and i_{R2}, and get

$$u_2 = \left(1 + \frac{R_2}{R_1}\right) u_1$$

It can be seen that the output voltage u_2 is controlled by the input voltage u_1, the circuit model is shown in Figure 3 - 1(a), and the transfer voltage ratio is $\mu = \left(1 + \frac{R_2}{R_1}\right)$.

(2) Voltage-Controlled Current Source (VCCS). The circuit of voltage-controlled current source is shown in Figure 3 - 4.

Using the characteristic 1 of the ideal operational amplifier, get $u_+ = u_- = u_1$, hence

$$i_R = \frac{u_1}{R_1}$$

Applying the characteristic 1 of the ideal operational amplifier, gives $i_2 = i_R = \frac{u_1}{R_1}$,

i_2 is controlled only by the input voltage u_1, and has nothing to do with the load resistance R_L (in fact, it is required that the resistance value of R_L is limited). The circuit model is shown in Figure 3 - 1(b), the transconductance is $g = \frac{i_2}{u_1} = \frac{1}{R_1}$.

(3) Current-Controlled Voltage Source (CCVS). The circuit of current-controlled voltage source is shown in Figure 3 - 5.

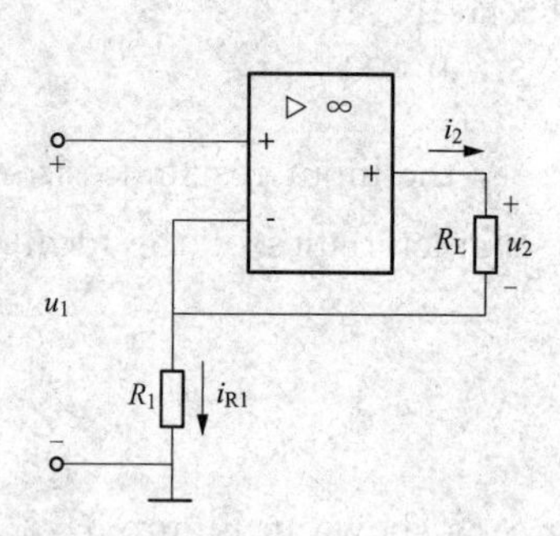

Figure 3 - 4 The Circuit of VCCS

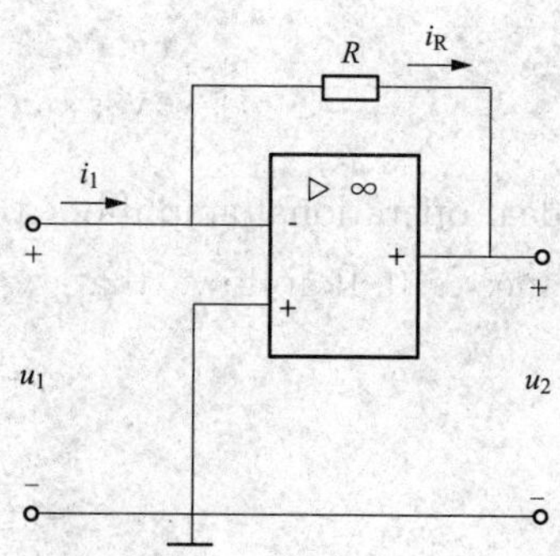

Figure 3 - 5 The Circuit of CCVS

Utilizing characteristic 1 of the ideal operational amplifier, yields $u_- = u_+ = 0$, $u_2 = Ri_R$ ①

Using characteristic 2 of operational amplifier, gives $i_R = i_1$ ②

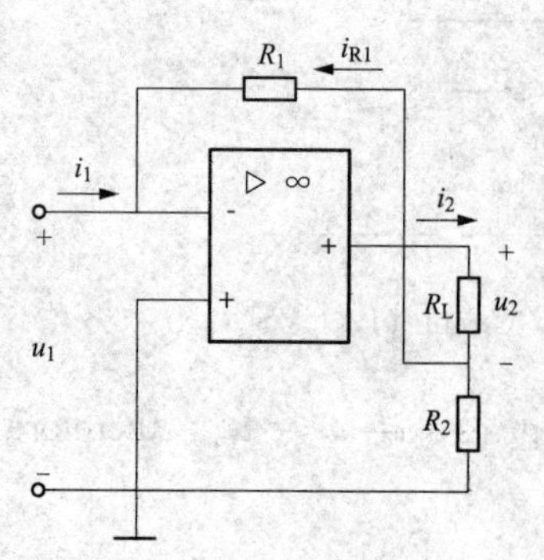

Figure 3 - 6 The Circuit of CCCS

Substituting equation ② into equation ①, get $u_2 = Ri_1$

The output voltage u_2 is controlled by the input current i_1. The current model is shown in Figure 3 - 1(c), and the transresistance is $r = \frac{u_2}{i_1} = R$.

(4) Current-Controlled Current Source (CCCS). The current-controlled current source circuit is shown in Figure 3 - 6.

Applying characteristic 1 of operational amplifier, gives $u_- = u_+ = 0$, $i_{R1} = \frac{R_2}{R_1 + R_2} i_2$ ③

Using characteristic 2 of the ideal operational amplifier, gives $i_{R1} = -i_1$ ④

Substituting equation ④ into equation ③, get $i_2 = -\left(1 + \frac{R_1}{R_2}\right) i_1$

i_2 is only controlled by the input current i_1, and has nothing to do with the load resistance R_L. The circuit model is shown in Figure 3 - 1(d), and the transfer current ratio is $\beta = \frac{i_2}{i_1} = -\left(1+\frac{R_1}{R_2}\right)$.

- **Equipment**

Equipment is shown in Table 3 - 1.

Table 3 - 1 Equipment

Equipment	Model and Specification	Quantity	Module
Constant Voltage Source	0～30V	1	NDG - 02
Constant Current Source	0～500mA	1	
DC Voltmeter	0～200V	1	NDG - 03
DC Ammeter	0～2000mA	1	
Experiment Circuit	Controlled Sources		NDG - 11
Resistor	1kΩ	1	NDG - 13
Adjustable Resistance	0～9999Ω	2	NDG - 06

- **Contents**

1. Test the Characteristics of VCCS

The experiment circuit is shown in Figure 3 - 1(b), where u_1 is the output of the constant voltage source, and the load resistor $R_L = 1\text{k}\Omega$ is connected to the two terminals of i_2.

(1) Test the Transfer Characteristics of VCCS $i_2 = f(u_1)$.

Adjust the output voltage u_1 (with the reading of the voltmeter as the standard) of the constant voltage source, measure the corresponding output current i_2 with an ammeter, and fill in Table 3 - 2.

Table 3 - 2 Data of the Transfer Characteristics of VCCS

u_1 (V)	0	0.5	1.0	1.5	2.0	2.5	3.0	3.5	4.0
i_2 (mA)									

(2) Test the Load Characteristics of VCCS $i_2 = f(R_L)$.

Keep $u_1 = 3\text{V}$ unchanged, the resistor box is used as the load R_L, adjust the resistor box, measure the corresponding output current i_2 with the ammeter, and fill in Table 3 - 3.

Table 3 - 3 Data of the Load Characteristics of VCCS

R_L(kΩ)	1	2	3	4	5	6	7	8	9
i_2(mA)									

2. Test the Characteristics of CCVS

The experiment circuit is shown in Figure 3 - 1(c), in which i_1 is the output of the constant current source, and the load resistor $R_L = 1\text{k}\Omega$ is connected to the two terminals of u_2.

(1) Test the Transfer Characteristics of CCVS $u_2 = f(i_1)$.

Adjust the output current i_1 (with the reading of the ammeter as the standard) of the constant current source, measure the output voltage u_2 with the voltmeter, and fill in Table 3-4.

Table 3-4 Data of the Transfer Characteristics of CCVS

I_1 (mA)	0.1	0.15	0.2	0.25	0.3	0.4
U_2 (V)						

(2) Test the Load Characteristics of CCVS $u_2 = f(R_L)$.

Keep $i_1 = 0.3$mA unchanged, the resistor box is used as the load R_L, adjust the box, measure the corresponding output voltage u_2 with the voltmeter, fill in Table 3-5.

Table 3-5 Data of the Load Characteristics of CCVS

R_L (kΩ)	1	2	3	4	5	6	7	8	9
u_2 (V)									

3. Test the Characteristics of VCVS

Voltage-controlled voltage source (VCVS) is comprised of a voltage controlled current source (VCCS) in series with a current controlled voltage source (CCVS). They are shown in Figure 3-1(b) and (c), connect the input port u_1 of Figure 3-1(b) to the output of the constant voltage source, connect the output port i_2 of Figure 3-1(b) to the input port i_1 of Figure 3-1(c), and connect the output port u_2 of Figure 3-1(c) to the load resistance $R_L = 1$kΩ.

(1) Test the Transfer Characteristics of VCVS $u_2 = f(u_1)$.

Adjust the output voltage u_1 (with the reading of the voltmeter as the standard) of the constant voltage source, measure the corresponding output voltage u_2 with the ammeter, and fill in Table 3-6.

Table 3-6 Data of the Transfer Characteristics of VCVS

u_1 (V)	0	0.5	1.0	1.5	2.0
u_2 (V)					

(2) Test the Load Characteristics of VCVS $u_2 = f(R_L)$.

Keep $u_1 = 1$V unchanged, the resistor box is used as the load R_L, adjust the resistor box, measure the corresponding output voltage u_2 with the voltmeter, and fill in Table 3-7.

Table 3-7 Data of the Load Characteristics of VCVS

R_L (kΩ)	1	2	3	4	5	6	7	8	9
u_2 (V)									

4. Test the Characteristics of CCCS

Current-controlled current source (CCCS) is comprised of a current controlled voltage source (CCVS) in series with a voltage controlled current source (VCCS). They are shown in Figure 3-1(c) and 3-1(b), connect the input port i_1 of Figure 3-1(c) to the output of the constant current source, connect the output port u_2 of Figure 3 - 1(c) to the input port u_1 of Figure 3 - 1(b), and connect the output port i_2 of Figure 3 - 1(b) to the load resistance $R_L = 1k\Omega$.

(1) Test the Transfer Characteristics of CCCS $i_2 = f(i_1)$.

Adjust the output current i_1 (with the reading of the ammeter as the standard) of the constant current source, measure the corresponding output current i_2 with the ammeter, fill in Table 3 - 8.

Table 3 - 8 Data of the Transfer Characteristics of CCCS

i_1 (mA)	0	0.05	0.10	0.15	0.20	0.25	0.30
i_2 (mA)							

(2) Test the Load Characteristics of CCCS.

Keep $i_1 = 0.3mA$ unchanged, the resistor box is used as the load R_L, adjust the box, measure the corresponding output current i_2 with the ammter, and fill in Table 3 - 9.

Table 3 - 9 Data of the Load Characteristics of CCCS

R_L (kΩ)	1	2	3	4	5	6	7	8	9
i_2 (mA)									

● **Notes**

1. The output terminal of an operational amplifier must not be grounded. The voltage of the input port should not be too high (less than 5V).

2. The load of the constant current source as a power supply should not be open-circuited.

● **Questions**

1. Will the output polarity of controlled source change if the polarity of the control variable reverses?

2. Are the control characteristics of controlled sources applicable for AC signals?

● **Experiment Report**

1. Draw the transfer characteristics curves and load characteristics curves of four types of controlled sources using the experiment data, calculate the corresponding transfer parameter μ, g, r and β.

2. Explain which parameters of the circuits affect the transfer parameter μ, g, r and β, and how to change μ, g, r and β.

实验4　电压源、电流源及其等效变换

一、实验目的

(1) 掌握建立电源模型的方法。

(2) 掌握电源外特性的测试方法。

(3) 加深对电压源和电流源特性的理解。

(4) 研究电源模型等效变换的条件。

二、实验原理

1. 电压源和电流源

电压源具有端电压保持恒定不变，而输出电流的大小由负载决定的特性。其外特性，即端电压U与输出电流I的关系$U=f(I)$是一条平行于I轴的直线。实验中使用的恒压源在规定的电流范围内，具有很小的内阻，可以将它视为一个电压源。

电流源具有输出电流保持恒定不变，而端电压的大小由负载决定的特性。其外特性，即输出电流I与端电压U的关系$I=f(U)$是一条平行于U轴的直线。实验中使用的恒流源在规定的电流范围内，具有极大的内阻，可以将它视为一个电流源。

2. 实际电压源和实际电流源

实际上任何电源内部都存在电阻，通常称为内阻。因而，实际电压源模型可以用一个内阻R_S和电压源U_S串联表示，其端电压U随输出电流I增大而降低。在实验中，可以用一个小阻值的电阻与恒压源相串联来模拟一个实际电压源。

实际电流源模型可用一个内阻R_S和电流源I_S并联表示，其输出电流I随端电压U增大而减小。在实验中，可以用一个大阻值的电阻与恒流源相并联来模拟一个实际电流源。

3. 实际电压源和实际电流源的等效变换

一个实际的电源，就其外部特性而言，既可以看成是一个电压源，又可以看成是一个电流源。若视为电压源，则可用一个电压源U_s与一个电阻R_S相串联表示；若视为电流源，则可用一个电流源I_S与一个电阻R_S相并联来表示。若它们向同样大小的负载供出同样大小的电流和端电压，则称这两个电源是等效的，即具有相同的外特性。

实际电压源与实际电流源等效变换的条件为：

(1) 取实际电压源与实际电流源的内阻均为R_S。

(2) 已知实际电压源的参数为U_S和R_S，则实际电流源的参数为$I_S=\frac{U_S}{R_S}$和R_S，若已知实际电流源的参数为I_s和R_S，则实际电压源的参数为$U_S=I_SR_S$和R_S。

三、实验设备

实验设备见表4-1。

四、实验内容

1. 测定电压源（恒压源）与实际电压源的外特性

实验电路如图4-1所示，图中的电源U_S用恒压源0～30V可调电压输出端，并将输出电压调到 6V，R_1 取 200Ω 的固定电阻，R_2 取 470Ω 的电位器。调节电位器令其阻值由大至小变化，将电流表、电压表的读数记入表 4－2 中。

表 4－1　　实 验 设 备

设备名称	型号与规格	数量	实验模块
恒压源	0～30V	1	NDG－02
直流电压表	0～200V	1	NDG－03
直流电流表	0～2000mA	1	
电阻	200Ω	1	NDG－13
	51Ω	1	
	1kΩ	1	
电位器	470Ω	1	

表 4－2　　电压源（恒压源）外特性数据

I（mA）							
U（V）							

在图 4－1 电路中，将电压源改成实际电压源，如图 4－2 所示，图中内阻 R_S 取 51Ω 的固定电阻，调节电位器 R_2，令其阻值由大至小变化，将电流表、电压表的读数记入表 4－3中。

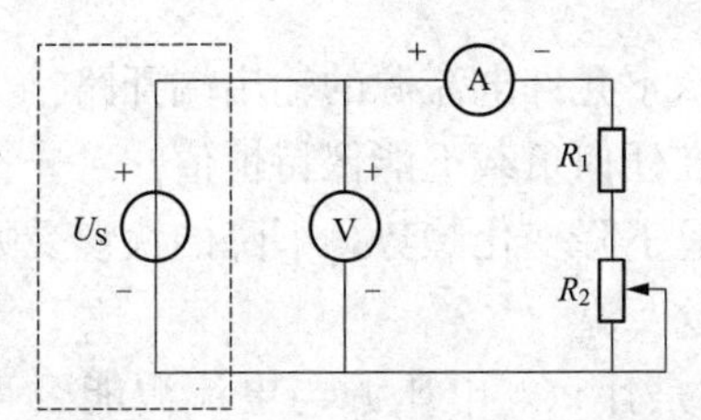

图 4－1　测定电压源的外特性

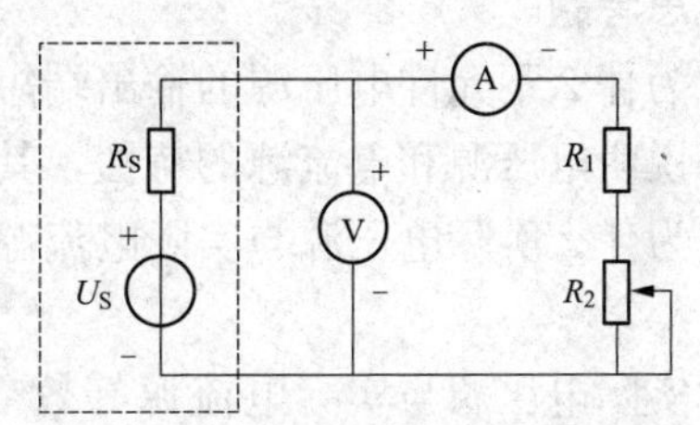

图 4－2　测定实际电压源的外特性

表 4－3　　实际电压源外特性数据

I（mA）							
U（V）							

2. 测定电流源（恒流源）与实际电流源的外特性

按图 4－3 接线，图中 I_S 为恒流源，调节其输出为 5mA（用电流表测量），R_2 取 470Ω 的电位器，在 R_S 分别为 1kΩ 和∞（开路）两种情况下，调节电位器 R_2，令其阻值由大至小变化，将电流表、电压表的读数记入自拟的数据表格中。

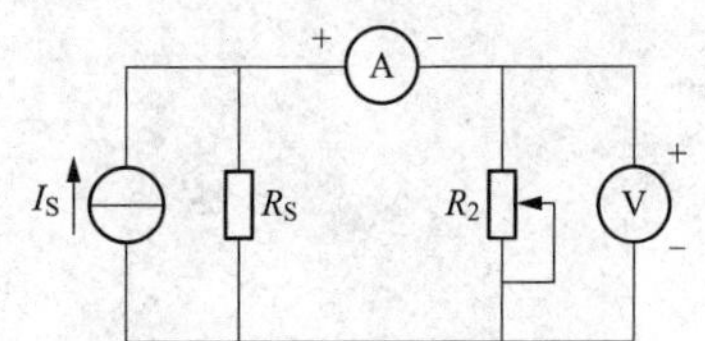

图 4－3　测定电流源的外特性

3. 研究电源等效变换的条件

按图 4－4 电路接线，其中图中的内阻 R_S 为 51Ω，负载电阻 R 均为 200Ω。

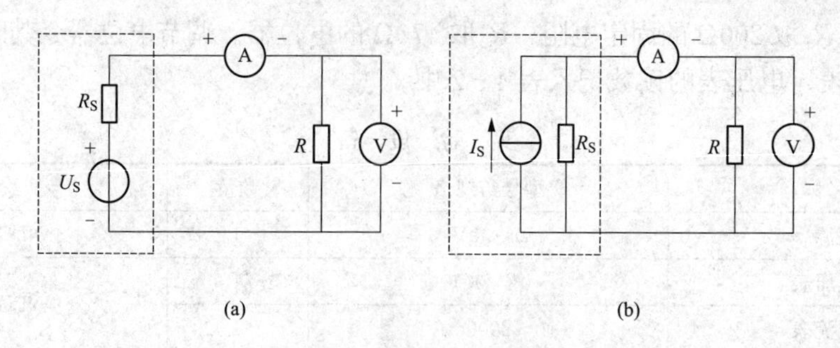

图 4－4 电源等效变换

(a) 实际电压源；(b) 实际电流源

在图 4－4(a) 电路中，U_S用恒压源 0～30V 可调电压输出端，并将输出电压调到 6V，记录电流表、电压表的读数。然后调节图 4－4(b) 电路中恒流源 I_S，令两表的读数与图 4－4(a) 的数值相等，记录 I_S的值，验证等效变换条件的正确性。

五、注意事项

(1) 在测电压源外特性时，不要忘记测空载（$I=0$）时的电压值；测电流源外特性时，不要忘记测短路（$U=0$）时的电流值，注意恒流源负载电压不可超过 20V，负载更不可开路。

(2) 换接线路时，必须关闭电源开关。

(3) 直流仪表的接入应注意极性与量程。

六、思考题

(1) 为什么不允许电压源的输出端短路？为什么不允许电流源的输出端开路？

(2) 说明电压源和电流源的特性，其输出是否在任何负载下能保持恒值？

(3) 为什么实际电压源与实际电流源的外特性呈下降变化趋势，下降的快慢受哪个参数影响？

(4) 实际电压源与实际电流源等效变换的条件是什么？电压源与电流源能否互相等效变换？

七、实验报告

(1) 根据实验数据绘出电源的四条外特性曲线，总结、归纳两类电源的特性。

(2) 根据实验结果，验证电源等效变换的条件。

Experiment 4 Verification of Voltage Source, Current Source and Their Equivalent Transformation

- **Objectives**

1. Learn how to model power sources.

2. Learn how to test the external characteristics of the power sources.

3. Acquire a deep understanding of the characteristics of the voltage source and the current source.

4. Understand the conditions of equivalent transformation of the power models.

- **Principles**

1. The Ideal Voltage Source and the Ideal Current Source

The characteristic of constant voltage source is that the terminal voltage is constant and the output current is determined by an ideal independent load. The external characteristic of voltage source, $U = f(I)$, is a straight line parallel to the I axis. The constant voltage source used in the experiment has a very low resistance value within the specified current range, which can be regarded as a voltage source.

The characteristic of a constant current source is that terminal current is constant and the terminal voltage is determined by the load. The external characteristic of current source, $I = f(U)$, is a straight line parallel to the U axis. The constant current source used in the experiment has a very high resistance value within the specified current range, which can be regarded as a current source.

2. The Actual Voltage Source and the Actual Current Source

In reality, any power source has an internal resistance. So the actual voltage source can be modeled as a small internal resistance R_S in series with a voltage source U_S. The terminal voltage U of an actual voltage source decreases with the increase of the output current I.

The actual current source can be modeled as a parallel connection of a high internal resistance R_S in parallel with a current source I_S. The output current I of an actual current source reduces with the increase of the terminal voltage U.

3. The Equivalent Transformation of an Actual Voltage Source and an Actual Current Source

An actual power source can be regarded as a voltage source or a current source in terms of its external characteristics. So the actual power source can be modeled as a series a resistance R_S in series with a voltage source U_S when it's regarded as a voltage source; and it can be modeled as a resistance R_S in parallel with a current source I_S when it's regarded as a current source. If two power sources deliver the same voltage and current to the same load, the two sources are said to be equivalent, in other words, they have same external characteristics.

The conditions of the equivalent transformation of the actual voltage source and actual current source are:

(1) The value of internal resistance of the actual voltage source is the same as that of the actual current source, say, R_S.

(2) If the parameters of the actual voltage source are known as U_s and R_S, then the parameters of the actual current source are $I_S = \frac{U_S}{R_S}$ and R_S.

If the parameters of the actual current source are known as I_S and R_S, then the parameters of the actual voltage source are $U_S = I_S R_S$ and R_S.

● Equipment

Equipment is shown in Table 4 - 1.

Table 4 - 1 Equipment

Equipment	Model or Specification	Quantity	Module
Constant Voltage Source	0～30V	1	NDG - 02
DC Voltmeter	0～200V	1	NDG - 03
DC Ammeter	0～2000mA	1	
Resistor	200Ω	1	NDG - 13
	51Ω	1	
	1kΩ	1	
Potentiometer	470Ω	1	

● Contents

1. Measure the External Characteristics of the Voltage source (Constant Voltage Source) and the Actual Voltage Source

The experiment circuit is shown in Figure 4 - 1. The output voltage of the constant voltage source U_S is adjusted to 6V. R_1 is a fixed resistor of 200Ω and R_2 is a potentiometer of 470Ω. Adjust the resistance of potentiometer from high value to low value, fill in Table 4 - 2 with the readings of the ammeter and the voltmeter.

Table 4 - 2 Data of the External Characteristic of the Voltage source (Constant Voltage Source)

I (mA)							
U (V)							

Change the voltage source for the actual voltage source, as shown in Figure 4 - 2, the internal resistance R_S is a fixed resistor of 51Ω. Adjust the resistance of potentiometer R_2 from high value to low value, fill in Table 4 - 3 with the readings of the ammeter and the voltmeter.

Table 4 - 3 Data of the External Characteristic of the Actual Voltage Source

I (mA)							
U (V)							

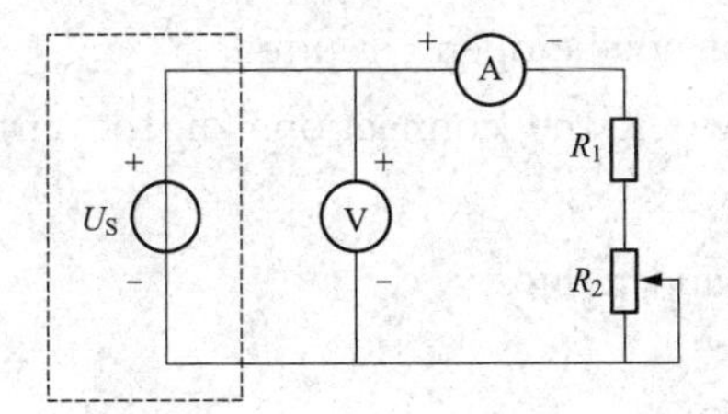

Figure 4－1 Measure the External Characteristics of the Voltage Source

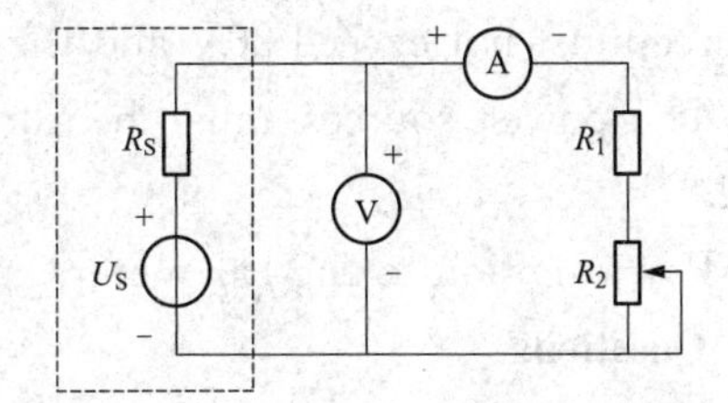

Figure 4－2 Measure the External Characteristics of the Actual Voltage Source

2. Measure the External Characteristics of the Current Source (Constant Current Source) and the Actual Current Source

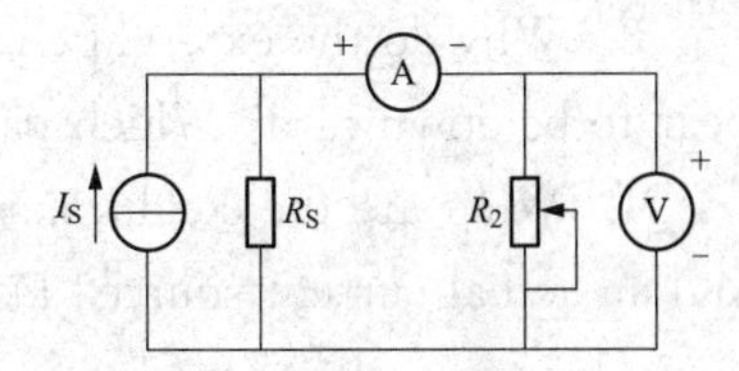

Figure 4－3 Measure the External Characteristics of the Current Source

Connect the circuit as shown in Figure 4－3. I_S is a constant current source. Adjust the output of I_S to 5mA (measured by the ammeter). R_2 is a potentiometer of 470Ω. Adjust the resistance of R_2 from high value to low value when R_S is 1kΩ and ∞ (open-circuit), fill in the self-made data table with the readings of the ammeter and the voltmeter.

3. Study the Conditions of Equivalent Transformation of Power Sources

Connect the circuits in Figure 4－4, separately, the internal resistances in Figure 4－4 are both 51Ω, and the load resistances are both 200Ω.

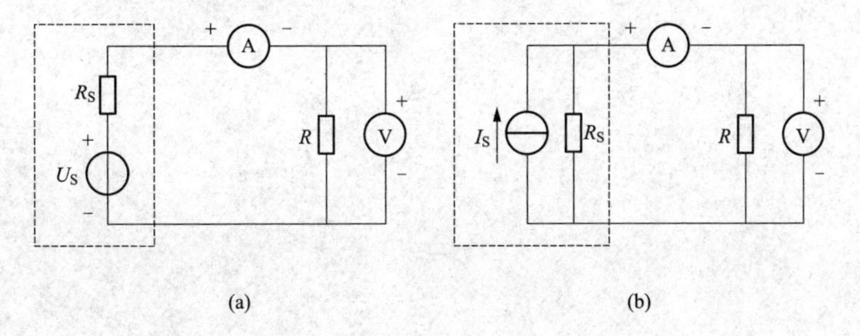

Figure 4－4 Equivalent Transformation of Power Sources

(a) Actual Voltage Source; (b) Actual Current Source

In the circuit of Figure 4－4(a), the power source U_S is the constant voltage source, whose output voltage is to 6V, and write down the readings of the ammeter and the voltmeter. Adjust the output of constant current source I_S in the Figure 4－4(b), set the readings of the two meters equal to the previous ones, write down the values of I_S, and verify the conditions of equivalent transformation.

● **Notes**

1. Do not forget to measure the no load voltage when the external characteristics of the voltage sources are measured and the short-circuit current when the external characteristics of the current sources are measured. The voltage across the load supplied by the constant cur-

rent source must not exceed 20V and the load must not be open-circuited.

2. The power sources must be turned off when the connections of the circuits are changed.

3. Pay attention to the DC meters' polarities and ranges.

● **Questions**

1. Why can the output terminals of voltage sources not be short-circuited? Why can the output terminals of current source not be open-circuited?

2. Do the output of voltage sources and current sources remain unchanged for any load? Explain both characteristics.

3. Why do the external characteristics of actual voltage sources and current sources appear to be downward? Which parameter affects the speed at which the curve descends?

4. What are the conditions of the equivalent transformation of an actual voltage source and an actual current source? Does an ideal voltage source and an ideal current source can be equivalent to each other?

● **Experiment Report**

1. Draw the four curves of external characteristics of the power sources, and summarize the characteristics of the two kinds of power sources.

2. Verify the conditions of equivalent transformation with the experiment results.

实验5　戴维南定理和诺顿定理的验证

一、实验目的

(1) 验证戴维南定理、诺顿定理的正确性，加深对定理的理解。

(2) 掌握测量有源二端网络等效参数的一般方法。

二、实验原理

1. 戴维南定理和诺顿定理

戴维南定理指出：一个含独立电源、线性电阻和受控源的二端网络（又称一端口网络），如图 5-1(a) 所示，总可以用一个电压源 U_S 和一个电阻 R_S 串联组成的实际电压源来代替，如图 5-1(b) 所示，其中电压源 U_S 等于这个有源二端网络的开路电压 U_{OC}，内阻 R_S 等于该网络中所有独立电源均置零（电压源短接，电流源开路）后的等效电阻 R_O。

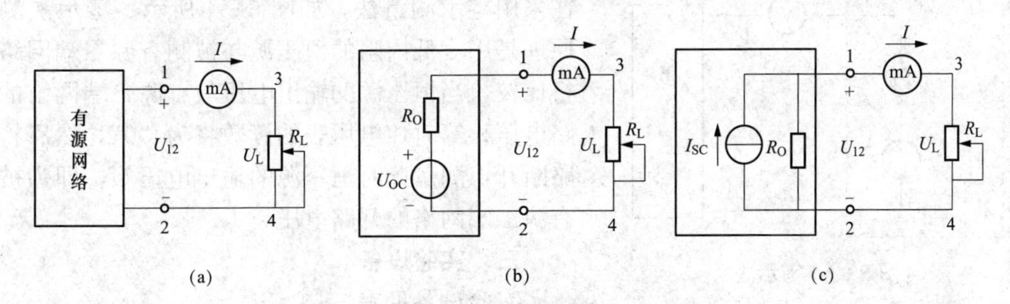

图 5-1　线性有源二端网络的等效置换

(a) 有源二端网络；(b) 等效电压源电路；(c) 等效电流源电路

诺顿定理指出：任何一个含独立电源、线性电阻和受控源的有源二端网络，如图 5-1(a) 所示，总可以用一个电流源 I_S 和一个电阻 R_S 并联组成的实际电流源来代替，如图 5-1(c) 所示，其中：电流源 I_S 等于这个有源二端网络的短路电源 I_{SC}，内阻 R_S 等于该网络中所有独立电源均置零（电压源短接，电流源开路）后的等效电阻 R_O。

U_S、R_S 和 I_S、R_S 称为有源二端网络的等效参数。

2. 有源二端网络等效参数的测量方法

(1) 开路电压、短路电流法。在有源二端网络输出端开路时，用电压表直接测其输出端的开路电压 U_{OC}，然后再将其输出端短路，测其短路电流 I_{SC}，则内阻为 $R_S=\dfrac{U_{OC}}{I_{SC}}$。

若有源二端网络的内阻值很低时，则不宜测其短路电流。

(2) 伏安法。用电压表、电流表测出有源二端网络的外特性曲线，如图 5-2 所示。开路电压为 U_{OC}，根据外特性曲线求出斜率 $\tan\varphi$，则内阻为 $R_S=\tan\varphi=\dfrac{\Delta U}{\Delta I}$。也可以先测量有源二端网络的开路电压 U_{OC}，以及额定电流 I_N 和对应的输出端额定电压 U_N，如图 5-2 所示，则内阻为 $R_S=\dfrac{U_{OC}-U_N}{I_N}$。

（3）半电压法。如图 5－3 所示，当负载电压为被测网络开路电压 U_{OC}一半时，负载电阻 R_L的大小（由电阻箱的读数确定）即为被测有源二端网络的等效内阻 R_S的数值。

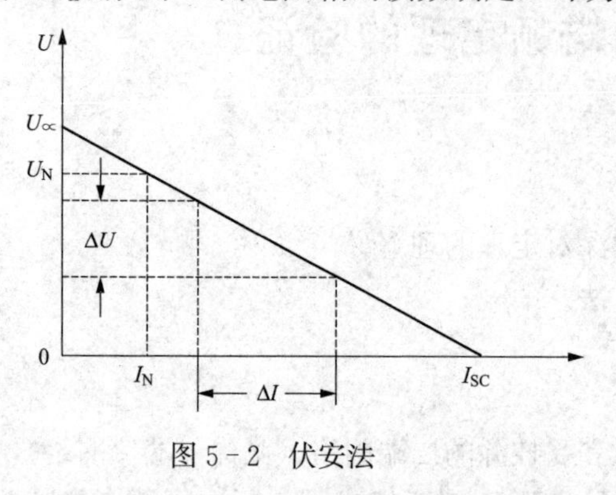

图 5－2 伏安法

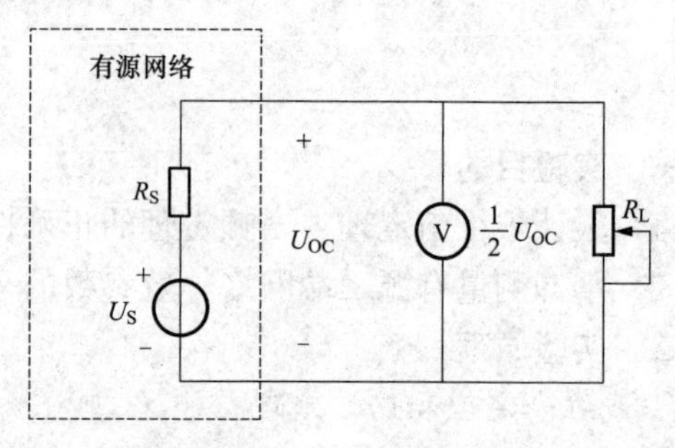

图 5－3 半电压法

（4）零示法。在测量具有高内阻有源二端网络的开路电压时，用电压表进行直接测量会造成较大的误差，为了消除电压表内阻的影响，往往采用零示测量法，如图 5－4 所示。零示法测量原理是用一低内阻的恒压源与被测有源二端网络进行比较，当恒压源的输出电压与有源二端网络的开路电压相等时，电压表的读数将为“0”，然后将电路断开，测量此时恒压源的输出电压 U，即为被测有源二端网络的开路电压。

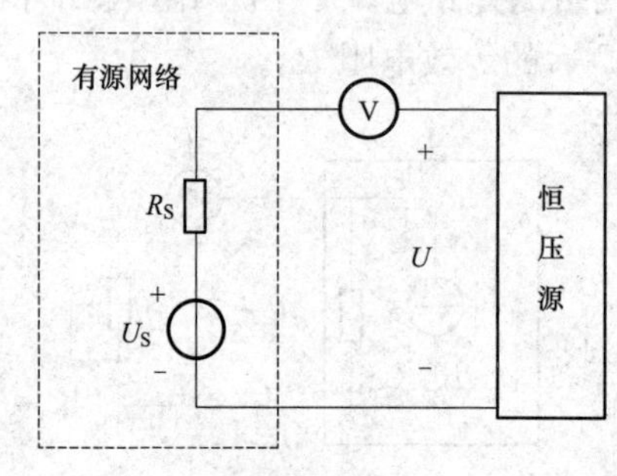

图 5－4 零示法

三、实验设备

实验设备见表 5－1。

表 5－1 实 验 设 备

设备名称	型号与规格	数量	实验模块
恒压源	0～30V	1	NDG－02
恒流源	0～500mA	1	
直流电压表	0～200V	1	NDG－03
直流电流表	0～2000mA	1	
实验电路	戴维南定理	1	NDG－12
电位器	1kΩ	1	NDG－12
可调电阻箱	0～9999Ω	2	NDG－06

四、实验内容

被测有源二端网络如图 5－5 所示。

1. 用开路电压、短路电流法测定 U_{OC}、I_{SC}和 R_S

在图 5－5 所示线路接入恒压源 U_S＝12V 和恒流源 I_S＝20mA。

在图 5－5 电路中，令端口 5、6 开路，用电压表测量开路电压 U_{OC}；令端口 5、6 短路，用电流表测量短路电流 I_{SC}，将数据记入表 5－2 中。计算内阻 R_s。

2. 测量有源二端网络的外特性

在图 5－5 电路中，在端口 5、6 处接入负载电阻 R_L（1kΩ 电位器），改变 R_L的阻值，逐点测量对应的电压U_L和电流I，将数据记入表5-3中。

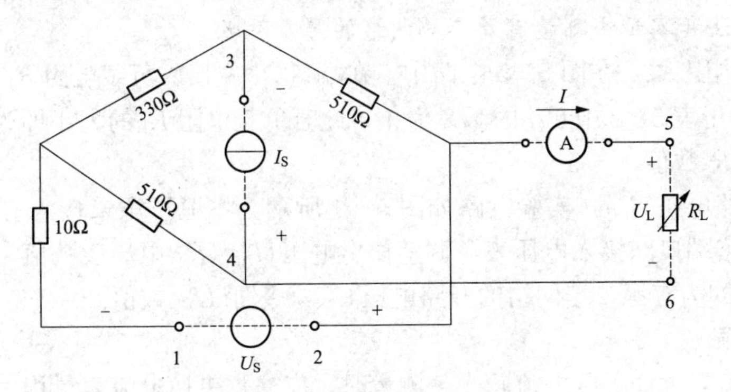

图5-5　被测有源二端网络

表5-2　被测有源二端网络的开路电压、短路电流及等效电阻

U_{OC} (V)	I_{SC} (mA)	$R_s=U_{OC}/I_{SC}$ (Ω)

表5-3　被测有源二端网络的外特性数据

U (V)										
I (mA)										

3. 测量等效电压源的外特性，验证戴维南定理

图5-1(b) 是图5-5有源二端网络的等效电压源电路，图中，电压源U_S用恒压源的输出端，调整到表5-2中的U_{OC}数值，内阻R_S按表5-2中计算出的R_S在电阻箱上选取并连接。负载R_L为1kΩ电位器，改变R_L的阻值，逐点测量对应的电压U_L和电流I，将数据记入表5-4中。

表5-4　等效电压源的外特性数据

U (V)										
I (mA)										

4. 测量等效电流源的外特性，验证诺顿定理

图5-1(c) 是图5-5有源二端网络的等效电流源电路，图中，电压源I_S用恒流源的输出端，恒流源调整到表5-2中的I_{SC}数值，内阻R_S按表5-2中计算出的R_S在电阻箱上选取并连接。负载R_L为1kΩ电位器，改变R_L的阻值，逐点测量对应的电压U_L和电流I，将数据记入表5-5中。

表5-5　等效电流源的外特性数据

U (V)										
I (mA)										

5. 测定有源二端网络等效电阻（又称入端电阻）的其他方法

将图5-5被测有源网络内的所有独立源置零（电流源I_S去掉，电压源U_S去掉并将端口1、2短接），然后用伏安法或直接用万用表的欧姆挡测定网络端口5、6处电阻，此即为被

测网络的等效内阻 R_{eq} 或称网络的入端电阻 R_{in}，记录该电阻值。

6. 用半电压法和零示法测量有源二端网络的等效参数

半电压法测内阻 R_S：在图 5－5 电路中，在端口 5、6 接入负载电阻 R_L，调节 R_L 直到负载两端电压 U_L 等于表 5－2 中 U_{OC} 的 1/2 为止，此时负载电阻 R_L 的大小即为有源网络的内阻 R_S 的数值。记录 R_S 数值。

零示法测开路电压 U_{OC}：实验电路如图 5－4 所示，将恒压源未接入有源网络内部的一路与被测有源网络端口间接入电压表，调整输出电压 U，观察电压表数值，当其等于零时输出电压 U 的数值即为有源二端网络的开路电压 U_{OC}，记录 U_{OC} 数值。

五、注意事项

（1）短路电流较大而负载电流较小，测量时，应注意更换电流表量程。

（2）注意实验电路中电压源与电流源的参考方向。

（3）独立源置零时不可将恒压源短接。

六、思考题

（1）在什么情况下不能直接测量有源二端网络的开路电压和短路电流？

（2）测量有源二端网络开路电压及等效内阻的几种方法有何优缺点？

七、实验报告

（1）根据表 5－3 的数据，绘出有源二端网络的外特性曲线，并根据外特性曲线计算出等效内阻 R_S。

（2）根据表 5－4 和表 5－5 的数据，绘出有源二端网络等效电路的外特性曲线，验证戴维南定理和诺顿定理的正确性。

（3）实验中用各种方法测得的 U_{OC} 和 R_S 是否相等？试分析其原因。

Experiment 5 Verification of Thevenin's Theorem and Norton's Theorems

● **Objectives**

1. Verify Thevenin's and Norton's theorems and have a better understanding of them.
2. Learn how to measure the equivalent parameters of a linear two-terminal circuit.

● **Principles**

1. Thevenin's and Norton's Theorems

Thevenin's theorem states that a linear two-terminal circuit (also known as one-port circuit) with one or more independent sources can be replaced by an equivalent circuit consisting of a voltage source U_S in series with a resistor R_S to the left of Figure 5 - 1(b) where U_S is the open-circuit voltage at the terminals and R_S is the input or equivalent resistance at the terminals when the independent sources are turn off. The linear two-terminal circuit and its Thevenin equivalent are to the left of Figure 5 - 1(a) and (b), respectively.

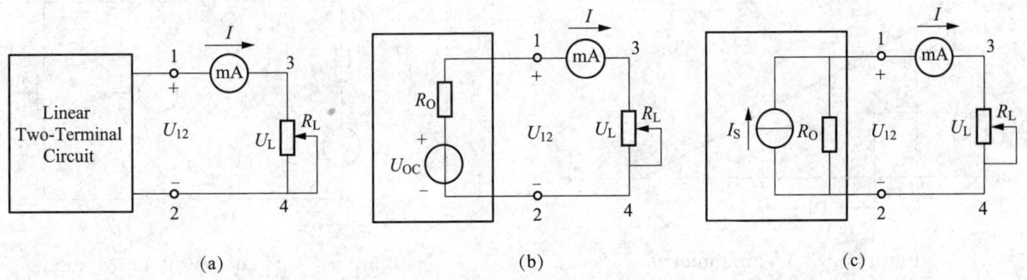

Figure 5 - 1 The Equivalent Replacements of a Linear Two-Terminal Circuit
(a) Two-Terminal Circuit; (b) The Equivalent Circuit with a Equivalent Voltage Source;
(c) The Equivalent Circuit with a Equivalent Current Source

Norton's theorem states a linear two-terminal circuit with one or more independent sources, can be replaced by an equivalent circuit consisting of a current source I_S in parallel with a resistor R_S, where I_S is the short-circuit current through the terminal and R_S is the input or equivalent resistance at the terminals when the independent sources are turn off. The linear two-terminal circuit and its Norton equivalent are to the left of Figure 5 - 1(a) and (c), respectively.

U_S, R_S and I_S, R_S are called the equivalent parameters of the linear two-terminal circuit.

2. The Measurement of Equivalent Parameters of the Linear Two-Terminal Circuit

(1) The Open-circuit Voltage and Short-circuit Current Method

Measure the open-circuit voltage U_{OC} with the voltmeter when the terminals of the linear two-terminal circuit are open-circuited, then make the terminals short-circuited, measure the short-circuit current I_{SC}, the internal resistance is $R_S = \dfrac{U_{OC}}{I_{SC}}$.

The short-circuit current should not be measured when the internal resistance of the linear two-terminal circuit is quite small.

(2) Voltammetry

Draw the external characteristic curve of the linear two-terminal circuit, as shown in Figure 5-2. The open-circuit voltage is U_{OC}, calculate the slope $\tan\varphi$ of the curve, the internal resistance is $R_S = \tan\varphi = \frac{\Delta U}{\Delta I}$.

The other way is to measure the open-circuit voltage U_{OC}, the rated current I_N and the rated voltage U_N at the output terminals corresponding to I_N, as shown in Figure 5-2, and the internal resistance is $R_S = \frac{U_{OC} - U_N}{I_N}$.

(3) Half-Voltage Method

As shown in Figure 5-3, when the voltage across the load is half the open-circuit voltage U_{OC} of the measured circuit, the load resistance determined by the reading of the resistor box is the equivalent internal resistance of the measured linear two-terminal circuit.

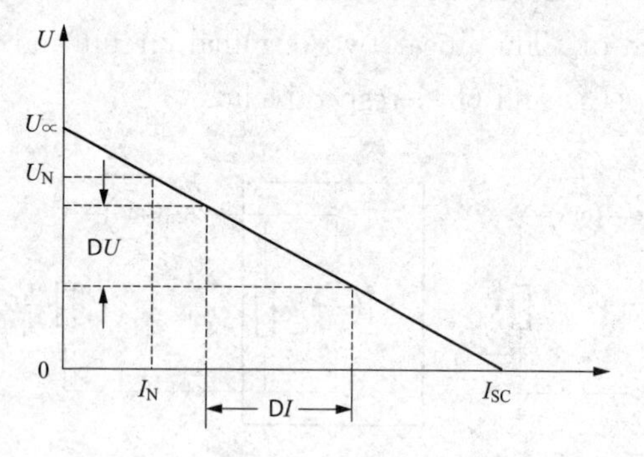

Figure 5-2 Voltammetry

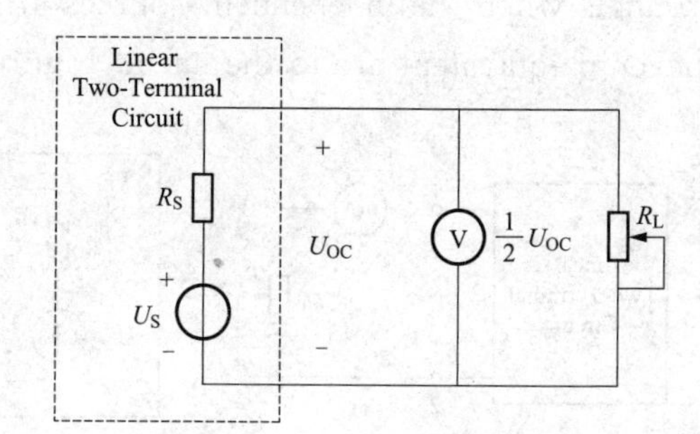

Figure 5-3 Half-Voltage Method

(4) Zero-Display Method

Measuring the open-circuit voltage of a linear two-terminal circuit with large internal resistance directly with a voltmeter will cause a big error. The zero-display measurement method is adopted to eliminate the influence of the internal resistance of the voltmeter, as shown in Figure 5-4. The principle of zero-display method is to compare the output voltage of a constant voltage source with small internal resistance and the open-circuit voltage of the measured active network. When the output voltage of the constant voltage source equals the open-circuit voltage of the measured active network, the reading of the voltmeter will be zero. In this case, disconnect the circuit, measure the output voltage U of the constant voltage source, which is the open-circuit voltage of measured active network.

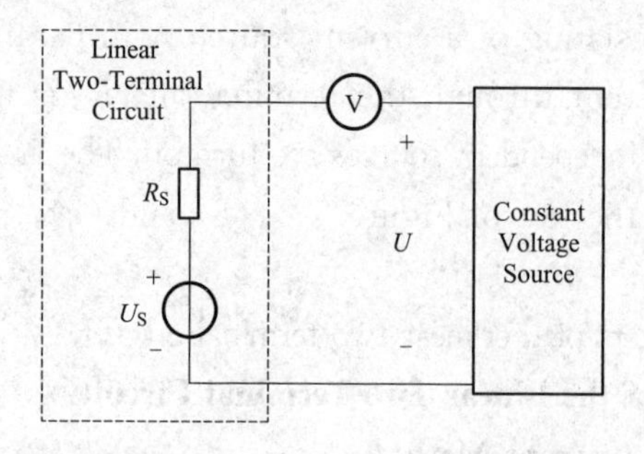

Figure 5-4 Zero-Display Method

● Equipment

Equipment is shown in Table 5 - 1.

Table 5 - 1 **Equipment**

Equipment	Model or Specification	Quantity	Module
Constant Voltage Source	0～30V	1	NDG - 02
Constant Current Source	0～500mA	1	
DC Voltmeter	0～200V	1	NDG - 03
DC Ammeter	0～2000mA	1	
Experiment Circuit	Thevenin's Theorem	1	NDG - 12
Potentiometer	1kΩ	1	NDG - 12
Adjustable Resistor	0～9999Ω	2	NDG - 06

● Contents

The measured linear two-terminal circuit is shown in Figure 5 - 5.

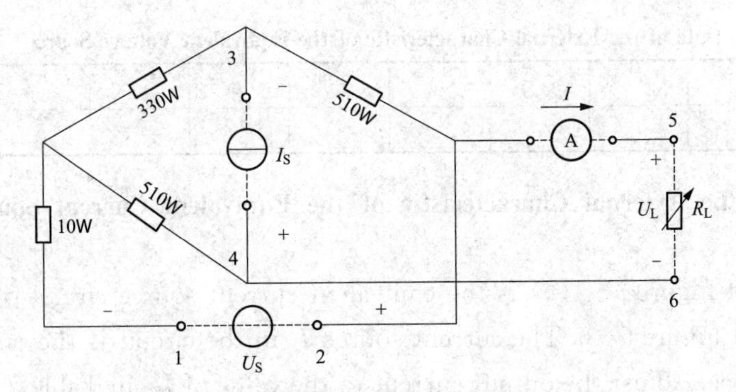

Figure 5 - 5 The Mearsured Linear Two-Terminal Circuit

1. Measure U_{OC}, I_{SC} and R_S with the Open-Circuit Voltage and Short-Circuit Current Method

Connect the constant voltage source U_S(=12V) and the constant current source I_S(=20mA) to the circuit shown in Figure 5 - 5.

In the circuit of Figure 5 - 5, measure the open-circuit voltage U_{OC} when terminals 5 and 6 are open-circuited; measure the short-circuit current I_{SC} when terminals 5 and 6 are short-circuited, fill in Table 5 - 2, calculate the internal resistance R_s.

Table 5 - 2 **the Open-Circuit Voltage, Short-Circuit Current and Internal Resistance of the Measured Two-Terminal Circuit**

Uoc (V)	Isc (mA)	Rs=Uoc/Isc (Ω)

2. Measure the External Characteristic of the Linear Two-Terminal Circuit

In the circuit of Figure 5 - 5, connect the load resistor R_L (1kΩ potentiometer) to the terminal 5 and 6, change the resistance of R_L, and measure the corresponding voltage U_L and the current I point by point, fill in Table 5 - 3.

Table 5 - 3 the External Characteristic of the Measured Two-Terminal Circuit

U (V)										
I (mA)										

3. Measure the External Characteristic of the Equivalent Voltage Source, and Verify Thevenin's Theorem

The circuit of Figure 5 - 1(b) is the equivalent voltage source circuit of the linear two-terminal circuit in Figure 5 - 5. The voltage source U_S in the circuit is the output of the constant voltage source. Adjust the output voltage to the value of U_{OC} in Table 5 - 2. The internal resistance R_S has already been calculated in Table 5 - 2. Choose and connect the resistors on the resistor box, make the sum of the resistance equal to R_S. The load R_L is the potentiometer of 1kΩ. Change the resistance of R_L and measure the corresponding voltage U_L and current I point by point. Fill in Table 5 - 4.

Table 5 - 4 Data of the External Characteristic of the Equivalent Voltage Source

U (V)										
I (mA)										

4. Measure the External Characteristic of the Equivalent Current Source, and Verify Norton's Theorem

The circuit of Figure 5 - 1(c) is the equivalent current source circuit of the linear two-terminal circuit in Figure 5 - 5. The current source I_S in the circuit is the output of the constant current source. Adjust the output current to the value of I_{SC} in Table 5 - 2. The internal resistance R_S has already been calculated in Table 5 - 2. Choose and connect the resistors on the resistor box, make the sum of the resistance equal to R_S. The load R_L is the potentiometer of 1 kΩ. Change the resistance of R_L and measure the corresponding voltage U_L and current I point-by-point. Fill in Table 5 - 5.

Table 5 - 5 Data of the External Characteristic of the Equivalent Current Source

U (V)										
I (mA)										

5. The Other Methods of Measuring the Equivalent Resistance (the Input Resistance) of the Linear Two-Terminal Circuit

Set the both independent power sources in the measured active network of Figure 5 - 5 equal to zero (remove the current source I_S and the voltage source U_S, connect the terminal 1 to terminal 2 with a wire). Measure the resistance at the terminal 5 and 6 by voltammetry, or measure the resistance with a multimeter. The resistance acquired is the equivalent internal resistance of the measured linear two-terminal circuit R_{eq} (also known as the input resistance of the network R_{in}), write this resistance down.

6. Measure the Equivalent Parameters of the Linear Two-Terminal Circuit by the Half-Voltage Method and the Zero-Display Method

(1) Measure the internal resistance R_S by the half-voltage method. In the circuit of Figure 5 - 5, connect the load resistor R_L to the terminals 5 and 6, and adjust R_L until the voltage U_L across the load is half as large as voltage U_{OC} in Table 5 - 2. At this point, the value of resistance R_L is that of the internal resistance R_S of the linear two-terminal circuit, note the value of R_S down.

(2) Measure the open-circuit voltage U_{OC} by the zero-display method. The experiment circuit is shown in Figure 5 - 4. Insert a voltmeter between the measured active circuit and the constant voltage source that is not inserted into the active circuit. Adjust the output voltage U while observe the reading of the voltmeter. When the reading is zero, the value of the output voltage U is the open-circuit voltage U_{OC} of the linear two-terminal circuit, note the value of U_{OC} down.

● **Notes**

1. When the short-circuit current is larger and the load current is smaller, in the measuring process, range of the ammeter must be changed.

2. Pay attention to the reference directions of the voltage source and the current source in the experiment circuit.

3. The constant voltage source must not be shorted when the independent power sources in the circuit are set to zero.

● **Questions**

1. Under what condition can the open-circuit voltage and the short-circuit current not be directly measured?

2. What are the advantages and disadvantages of the methods of measuring the open-circuit voltage and the internal resistance of the linear two-terminal circuit?

● **Experiment Report**

1. Draw the external characteristic curve of the linear two-terminal circuit using the data of Table 5 - 3, and calculate the equivalent internal resistance R_S from the curve.

2. Draw the external characteristic curve of the linear two-terminal circuit employing the data of the Table 5 - 4 and Table 5 - 5, and verify Thevenin's theorem and Norton's theorem.

3. Are the U_{OC} and R_S measured by various methods equal? Try to analyze the reasons.

实验6 互易定理的验证

一、实验目的

(1) 验证互易定理。

(2) 进一步熟悉直流电流表、直流电压表及电压源和电流源的使用方法。

(3) 了解仪表误差对测量结果的影响。

二、实验原理

互易定理是线性电路的一个重要性质。互易定理指出，对于一个仅含线性电阻且只有一个激励的电路，在保持电路将独立电源置零后电路拓扑结构不变的条件下，激励和响应互换位置后，响应与激励的比值保持不变。

互易定理有三种形式。图6-1是互易定理的第一种形式，电路中只有一个独立电压源作为激励，如图6-1(a) 所示，电压源U_S接在电路中的1-1′端，将2-2′端短路，电流I_1是电压源U_S所产生的响应。则将电压源U_S移至2-2′端，而将1-1′端短路如图6-1(b) 所示，那么有$I_2=I_1$。

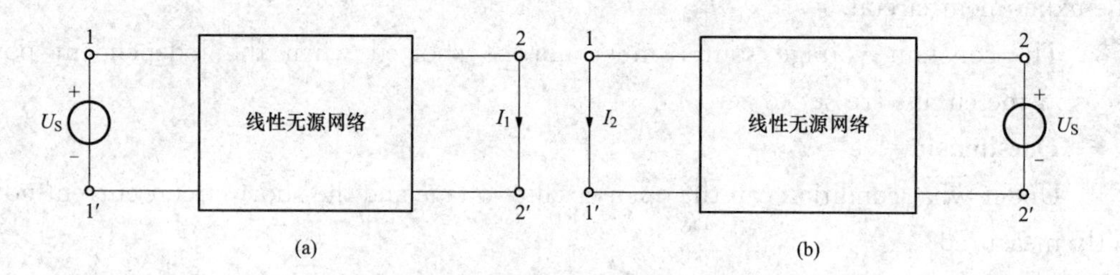

图6-1 互易定理的第一种形式

图6-2是互易定理的第二种形式，电路中只有一个独立电流源作为激励，如图6-2(a) 所示，电流源I_S接在电路中的3-3′端，将4-4′端开路，电路端口电压U_1是电流源所产生的响应。则将电流源I_S移至4-4′端，而将3-3′端开路如图6-2(b) 所示，那么有$U_2=U_1$。

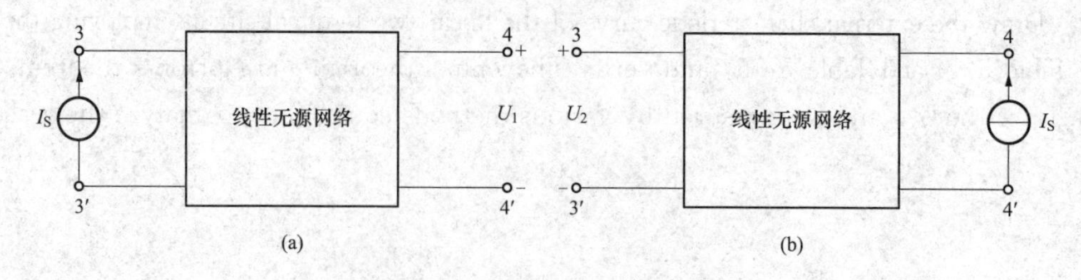

图6-2 互易定理的第二种形式

图6-3是互易定理的第三种形式，电流源I_S接在图6-3(a) 中的6-5′端，6-6′端短路电流I_3是电流源I_S所产生的响应，则若将电流源I_S改为一个数值上相等的电压源U_S移至6-6′端，而将6-5′端开路如图6-3(b) 所示，那么有$U_3=I_3$。

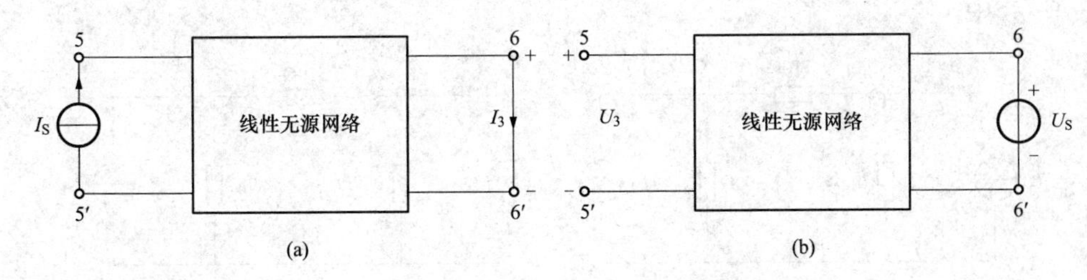

图 6-3 互易定理的第三种形式

三、实验设备

实验设备见表 6-1。

表 6-1 实 验 设 备

设备名称	型号与规格	数量	实验模块
恒压源	0～30V	1	NDG-02
恒流源	0～500mA	1	
直流电压表	0～200V	1	NDG-03
直流电流表	0～2000mA	1	
电阻	200Ω	1	NDG-13
	300Ω	1	
	510Ω	1	

四、实验内容

（1）按图 6-4 连接电路，激励源 U_{11} 由恒压源提供，首先将激励源接在 200Ω 的支路上，测量其在 300Ω 支路中的电流响应 I_{11}；再将激励源 U_{11} 移至 300Ω 的支路上，测量其在 200Ω 支路上的电流响应 I_{12}。激励源分别取 4 组不同的电压值，记录不同激励电压值下的 I_{11} 和 I_{12} 电流值于表 6-2 中。

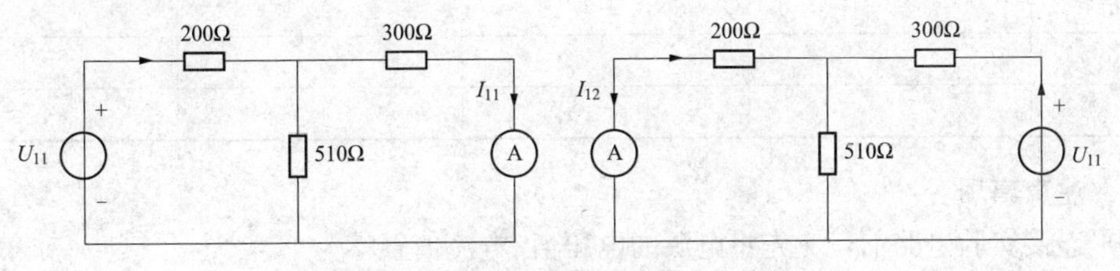

图 6-4 互易定理第一种形式的验证

表 6-2 互易定理第一种形式的验证数据

U_{11}（V）				
I_{11}（mA）				
I_{12}（mA）				

（2）按图 6-5 连接电路，激励源 I_{21} 由恒流源提供，先将激励源接在 200Ω 电阻上，测量其在 300Ω 电阻上的电压响应 U_{21}；再将激励源 I_{21} 接在 300Ω 电阻上，测量其在 200Ω 电阻上的电压响应 U_{22}。激励源分别取 4 组不同的电流值，记录不同电流值下的 U_{21} 和 U_{22} 的电压

值于表 6－3 中。

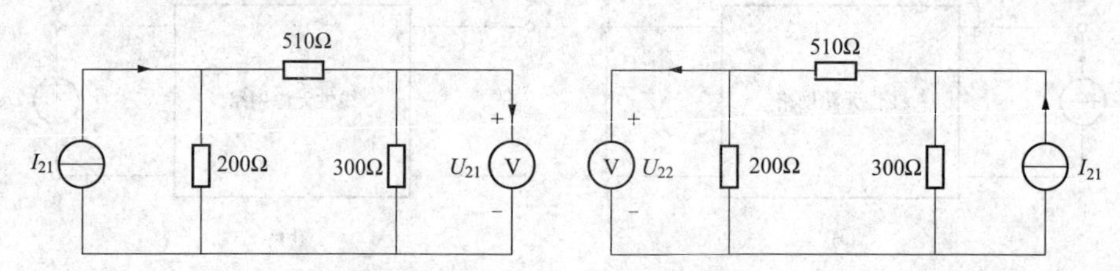

图 6－5 互易定理第二种形式的验证

表 6－3 **互易定理第二种形式的验证数据**

I_{21}（mA）				
U_{21}（V）				
U_{22}（V）				

（3）按图 6－6 连接电路，两组电路中的激励源 I_{31} 与 U_{31} 在数值上相等。先将激励电流源 I_{31} 接在 200Ω 支路上，测量其在 300Ω 支路上的电流响应 I_{32}；再将激励源改为电压源 U_{31} 接在 300Ω 支路上，测量其在 200Ω 支路上的开路电压 U_{32}。取 4 组不同的激励源的值，分别记录 I_{32} 和 U_{32} 的值于表 6－4 中。

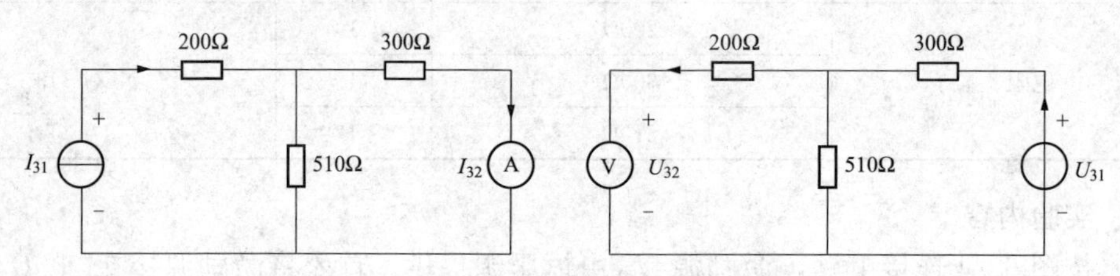

图 6－6 互易定理第三种形式的验证

表 6－4 **互易定理第三种形式的验证数据**

I_{31}（mA）				
I_{32}（mA）				
U_{31}（V）				
U_{32}（V）				

五、注意事项

（1）连接实验电路前，应先将电源的电压值/电流值置零。

（2）谨防恒压源输出端短路。

（3）注意测量数据的参考方向和实际方向。

六、实验报告

（1）结合电路图及电路参数，计算实验电路中被测数值。

（2）根据各表格所测数据，验证互易定理。

Experiment 6　Verification of the Reciprocity Theorem

● **Objectives**

1. Verify the reciprocity theorem.

2. Be more familiar with the use of DC ammeter, DC voltmeter, the voltage source and current source.

3. Understand the influence of the instrument errors on the measurement results.

● **Principles**

The reciprocity theorem is an important property of linear circuits. Reciprocity means for a circuit which only contains linear resistors and one excitation source, if the topology of the circuit is unchanged after the independent power source was put to zero, the ratio of response and excitation is also unchanged after their positions were swapped.

There are three forms of the reciprocity theorems. Figure 6 - 1 shows the first form of the theorem. An independent voltage source U_S as the excitation is connected to port 1 - 1′ with port 2 - 2′ short-circuited as shown in Figure 6 - 1(a), and the short-current I_1 is the response of U_S. When the voltage source U_S is connected to port 2 - 2′, with port 1 - 1′ is short-circuited, as shown in Figure 6 - 1(b), we get $I_2 = I_1$.

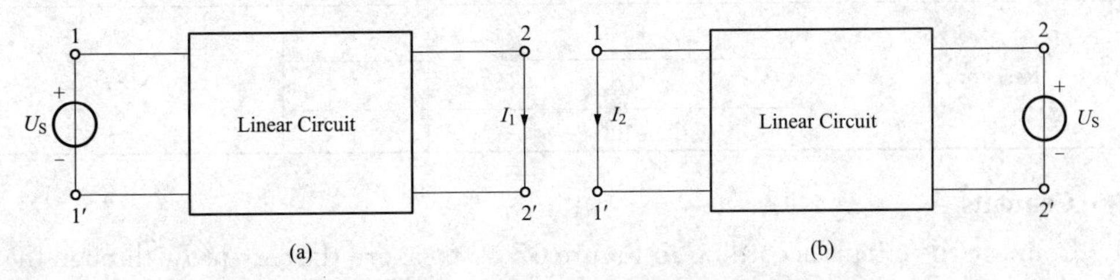

Figure 6 - 1　The First Form of the Reciprocity Theorem

Figure 6 - 2 shows the second form of the reciprocity theorem. An independent current source I_S as a excitation is connected to port 3 - 3′, with port 4 - 4′ open-circuited as shown in Figure 6 - 2(a), and the voltage U_2 is the response of I_S. When the current source I_S is connected to port 4 - 4′ with port 3 - 3′ open circuited as shown in Figure 6 - 2(b), we get $U_2 = U_1$.

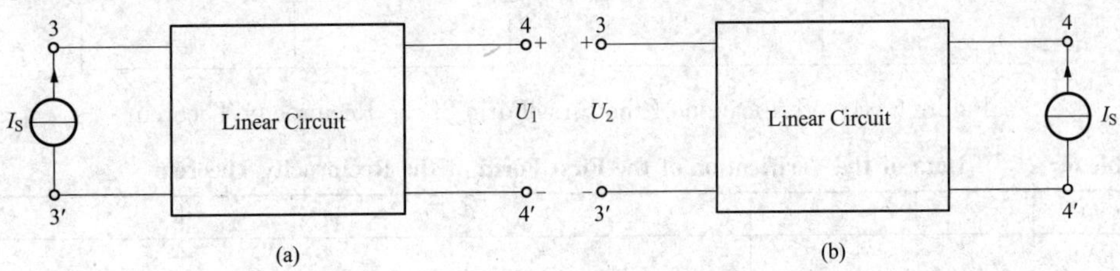

Figure 6 - 2　The Second Form of the Reciprocity Theroem

Figure 6 - 3 shows the third form of the reciprocity theorem. An independent current source I_S is connected to port 5 - 5′, with port 6 - 6′ short-circuited as shown in Figure 6 - 3(a), and the short circuit current I_3 is the response of I_S. When an independent voltage source U_S with equal value to I_S in value is connected to part 6 - 6′ with port 5 - 5′ open-circuited as shown in Figure 6 - 3(b), we get $U_3 = I_3$.

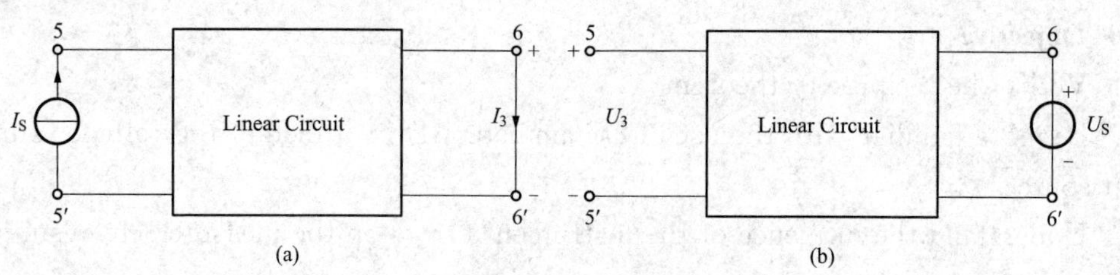

Figure 6 - 3 The Third Form of the Reciprocity Theroem

● Equipment

Equipment is shown in Table 6 - 1.

Table 6 - 1 Equipment

Equipment	Model or Specification	Quantity	Module
Constant Voltage Source	0～30V	1	NDG - 02
Constant Current Source	0～500mA	1	
DC Voltmeter	0～200V	1	NDG - 03
DC Ammeter	0～2000mA	1	
Resistor	200Ω	1	NDG - 13
	300Ω	1	
	510Ω	1	

● Contents

1. Connect the circuit according to Figure 6 - 4, measure the current I_{11} through the 300Ω resistor. Then connect U_{11} to the 300Ω branch, and measure the current I_{12} through the 200Ω resistor. When the voltage source U_{11} takes four different values, write down the corresponding current I_{11} and I_{12}, and fill in Table 6 - 2.

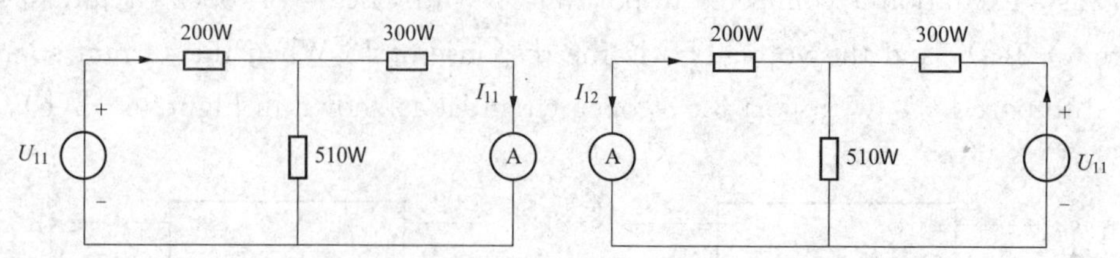

Figure 6 - 4 Verification of the First Form of the Reciprocity Theorem

Table 6 - 2 Data of the Verification of the First Form of the Reciprocity Theorem

U_{11} (V)				
I_{11} (mA)				
I_{12} (mA)				

2. Connect the circuit according to Figure 6 - 5, and measure the voltage U_{21} across the 300Ω resistor. Then connect U_{11} to the 300Ω resistor, and measure the voltage U_{22} across the 200Ω resistor. When the current source I_{21} takes four different values, write down the corresponding voltage U_{21} and U_{22}, and fill in Table 6 - 3.

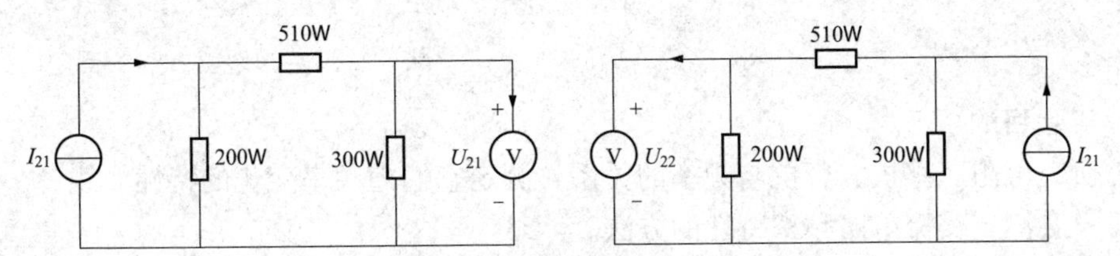

Figure 6 - 5 Verification of the Second Form of the Reciprocity Theorem

Table 6 - 3 Data of the Verification of the Second Form of the Reciprocity Theorem

I_{21} (mA)				
U_{21} (V)				
U_{22} (V)				

3. Connect the circuit according to Figure 6 - 6, where the current source I_{31} and voltage source U_{31} are equal in value. Measure the current I_{32} through the 300Ω resistor as shown in Figure 6 - 6(a), and the voltage U_{32} across the 200Ω resistor as shown in Figure 6 - 6(b). When I_{31} and U_{31} takes four different values, write down U_{32} and I_{32} for the different excitations, and fill in Table 6 - 4.

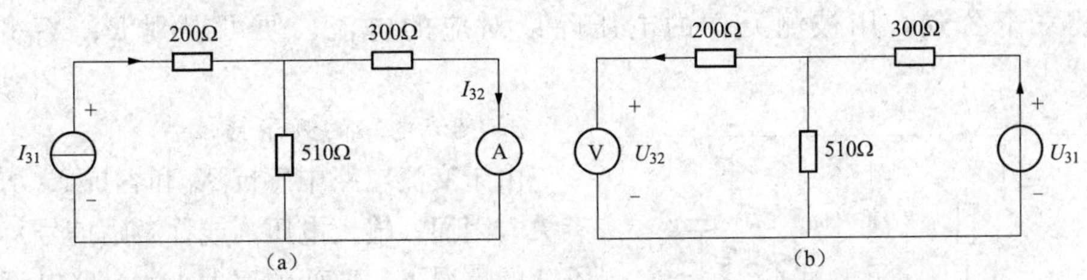

Figure 6 - 6 Verification of the Third Form of the Reciprocity Theorem

Table 6 - 4 Data of the Verification of the Third Form of the Reciprocity Throrem

I_{31} (mA)				
I_{32} (mA)				
U_{31} (V)				
U_{32} (V)				

- **Notes**

1. Power supplies must be turned off before connecting the experiment circuits.
2. Beware of short-circuiting of the constant voltage source.
3. Pay attention to the reference directions and the actual directions of the measured data.

- **Experiment Report**

1. Calculate the measured data in the experiment circuits according to the circuit diagram and the circuit parameters.
2. Verify the reciprocity theorem according to the data in the tables.

实验7 *R*、*L*、*C*元件与高通、低通、带通滤波器的频率特性

一、实验目的

(1) 研究电阻、感抗、容抗与频率的关系，测定它们随频率变化的特性曲线。

(2) 学会测定交流电路频率特性的方法。

(3) 了解滤波器的原理和基本电路。

二、实验原理

1. 单个元件阻抗与频率的关系

对于电阻元件，根据 $\frac{\dot{U}_R}{\dot{I}_R}=R\angle 0^\circ$，其中 $\frac{u_R}{i_R}=R$，电阻 R 与频率无关。

对于电感元件，根据 $\frac{\dot{U}_L}{\dot{I}_L}=\mathrm{j}X_L$，其中 $\frac{u_L}{i_L}=X_L=2\pi fL$，感抗 X_L 与频率成正比。

对于电容元件，根据 $\frac{\dot{U}_C}{\dot{I}_C}=-\mathrm{j}X_C$，其中 $\frac{u_C}{i_C}=X_C=\frac{1}{2\pi fC}$，容抗 X_C 与频率成反比。

测量元件阻抗频率特性的电路如图7-1所示，图中 r 是提供测量回路电流用的标准电阻，流过被测元件的电流（i_R、i_L、i_C）可由 r 两端的电压 u_r 除以 r 阻值所得，又根据上述三个公式，用被测元件的电压除以对应的电流，便可得到 R、X_L 和 X_C 的数值。

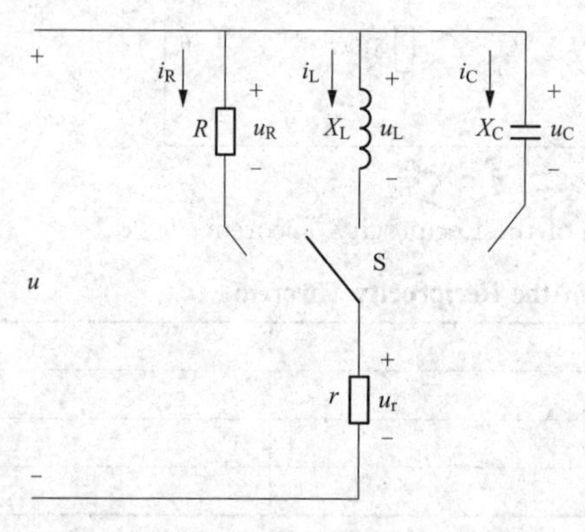

图7-1 测量元件阻抗频率特性电路

2. 交流电路的频率特性

由于交流电路中感抗 X_L 和容抗 X_C 均与频率有关，因而，输入电压（或称激励信号）在大小不变的情况下，改变频率大小，电路电流和各元件电压（或称响应信号）也会发生变化。这种电路响应随激励频率变化的特性称为频率特性。

若电路的激励信号为 $E_x(\mathrm{j}\omega)$，响应信号为 $R_e(\mathrm{j}\omega)$，则频率特性函数为

$$N(\mathrm{j}\omega)=\frac{R_e(\mathrm{j}\omega)}{E_x(\mathrm{j}\omega)}=A(\omega)\angle\varphi(\omega)$$

式中：$A(\omega)$ 为响应信号与激励信号的大小之比，是 ω 的函数，称为幅频特性；$\varphi(\omega)$ 为响应信号与激励信号的相位差角，也是 ω 的函数，称为相频特性。

在本实验中，研究几个典型电路的幅频特性，如图7-2所示。其中，图7-2(a) 在高频时有响应（即有输出），称为高通滤波器。图7-2(b) 在低频时有响应（即有输出），称为低通滤波器，图中对应 $A=0.707$ 的频率 f_C 称为截止频率，在本实验中用 RC 网络组成的高通滤波器和低通滤波器，它们的截止频率 f_C 均为 $1/2\pi RC$。图7-2(c) 在一个频带范围内有响应（即有输出），称为带通滤波器，图中 f_{C1} 称为下限截止频率，f_{C2} 称为上限截止频率，

通频带 $BW=f_{C2}-f_{C1}$。

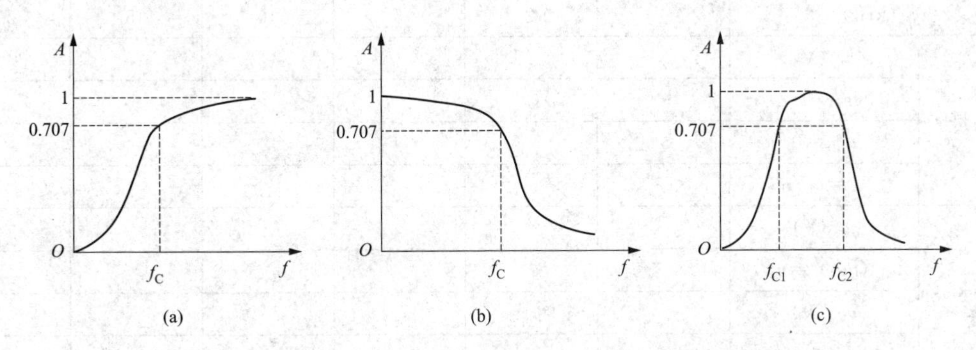

图7-2　三种典型滤波器的幅频特性

(a) 高通滤波器；(b) 低通滤波器；(c) 带通滤波器

三、实验设备

实验设备见表7-1。

表7-1　　**实　验　设　备**

设备名称	型号与规格	数量	实验模块
双踪示波器	GDS-1102A-U	1	
信号发生器	DG1022U	1	
电阻	300Ω	1	NDG-13
电阻	1kΩ	1	
电阻	2kΩ	1	
电感	10mH	1	
电容	0.01μF	1	

四、实验内容

1. 测量 R、L、C 元件的阻抗频率特性

实验电路如图7-1所示，图中：$r=300\Omega$，$R=1k\Omega$，$L=10mH$，$C=0.01\mu F$。选择信号源正弦波输出作为输入电压 u，调节信号源输出电压幅值，并用示波器测量，使输入电压 u 的有效值 $U=2V$，并保持不变。

用导线分别接通 R、L、C 三个元件，调节信号源的输出频率，从1kHz逐渐增至20kHz，用示波器分别测量 u_R、u_L、u_C 和 u_r，将实验数据记入表7-2中。并通过计算得到各频率点的 R、X_L 和 X_C。

表7-2　　**R、L、C元件的阻抗频率特性实验数据**

f（kHz）		1	2	5	10	15	20
R（kΩ）	u_r（V）						
	i_R（mA）$=u_r/r$						
	u_R（V）						
	$R=u_R/i_R$						

续表

f（kHz）		1	2	5	10	15	20
X_L（kΩ）	u_r（V）						
	i_L（mA）$=u_r/r$						
	u_L（V）						
	$X_L=u_L/i_L$						
X_C（kΩ）	u_r（V）						
	i_C（mA）$=u_r/r$						
	u_C（V）						
	$X_C=u_C/i_C$						

2. 高通滤波器频率特性

实验电路如图 7－3 所示，图中：$R=2\text{k}\Omega$，$C=0.01\mu\text{F}$。用信号源输出正弦波电压作为电路的激励信号（即输入电压）u_i，调节信号源正弦波输出电压幅值，并用示波器测量，使激励信号 u_i的有效值 $U_i=2\text{V}$，并保持不变。调节信号源的输出频率，从 1kHz 逐渐增至 20kHz，用示波器测量响应信号（即输出电压）u_R，将实验数据记入表 7－3 中。

表 7－3　　频率特性实验数据

f（kHz）	1	3	6	8	10	15	20
u_R（V）							
u_C（V）							
u_O（V）							

3. 低通滤波器频率特性

实验电路和步骤同实验 2，只是响应信号（即输出电压）取自电容两端电压 u_C，将实验数据记入表 7－3 中。

4. 带通滤波器频率特性

实验电路如图 7－4 所示，图中：$R=1\text{k}\Omega$，$L=10\text{mH}$，$C=0.1\mu\text{F}$。实验步骤同实验 2，响应信号（即输出电压）取自电阻两端电压 u_O，将实验数据记入表 7－3 中。

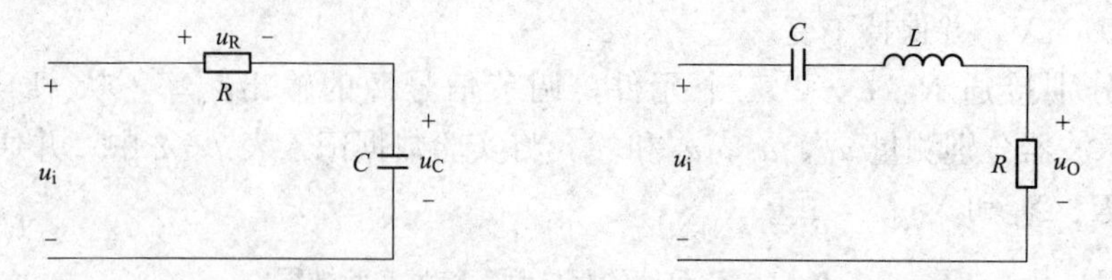

图 7－3　高通滤波器实验电路　　　图 7－4　低通滤波器和带通滤波器实验电路

五、注意事项

注意示波器的正确使用方法。

六、思考题

（1）如何用示波器测量电阻 R、感抗 X_L和容抗 X_C？它们的大小和频率有何关系？

（2）什么是频率特性？高通滤波器、低通滤波器和带通滤波器的幅频特性有何特点？如何测量？

七、实验报告

（1）根据实验数据，在方格纸上绘制 R、L、C 三个元件的阻抗频率特性曲线，并总结其特性。

（2）根据实验数据，在方格纸上绘制高通、低通、带通滤波器的频率特性曲线，并总结其特性。

Experiment 7 Frequency Characteristics of *R*, *L*, *C*, the High-Pass Filter, the Low-Pass Filter and the Band-Pass Filter

● Objectives

1. Study the relations between resistance, inductive reactance, capacitive reactance and frequency, measure the variation curves of these three parameters changing with frequency.
2. Learn how to measure frequency response of AC circuit.
3. Understand the principle of the filters and their basic circuits.

● Principles

1. The Relations between the Reactances of a Circuit Elements and Frequencies

For a resistor, since $R=\dfrac{U_R}{I_R}$, the resistance R is irrelevant to frequency.

For a inductor, since $X_L=\omega L=2\pi fL$, X_L is proportional to frequency.

For a capacitor, since $X_C=\dfrac{1}{\omega C}=\dfrac{1}{2\pi fC}$, X_C is inversely with the frequency.

The circuit for measuring the impedance frequency characteristic of circuit elements is shown in Figure 7 - 1, r is a standard resistor for measuring the current. The current through the measured elements (i_R, i_L, i_C) can be calculated by dividing the voltage u_r across the resistor r by the resistance of r, and R, X_L and X_C can be calculated by dividing the voltage across these elements by the current.

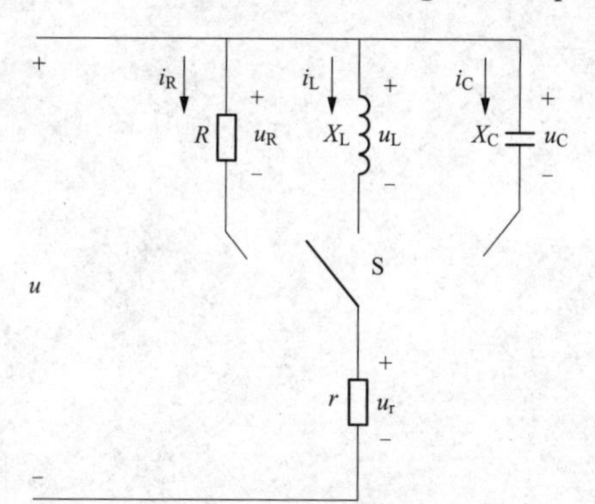

Figure 7 - 1 The Circuit for Measuriy the Impedance Frequency Character istics of Circuit Elements

2. The Frequency Characteristic of AC Circuit

In AC circuits, the inductanceX_L and capacitance X_C are related to frequency, keep the input voltage (or called excitation signal) constant, the current in the circuit and the voltages across elements (or called response signals) vary with the frequency of the input voltage. This characteristic that responses in the circuit vary with the excitation frequency is called frequency response.

If the excitation signal in the circuit is $E_x(j\omega)$, the response signal is $R_e(j\omega)$, the frequency response function is

$$N(j\omega)=\frac{R_e(j\omega)}{E_x(j\omega)}=A(\omega)\angle\varphi(\omega)$$

In the function, $A(\omega)$, a function of ω, is the ratio of the response signal to the excitation signal, called amplitude frequency characteristic; $\varphi(\omega)$, a function of ω too, is the phase difference angle between the response signal and the excitation signal, called phase frequency characteristic.

The amplitude frequency characteristics of several typical circuits, which are shown in Figure 7-2, are studied in this experiment. In this Figure 7-2(a) has response (output) at high frequency, called high-pass filter. Figure 7-2(b) has response at low frequency, called low-pass filter, the frequency f_C corresponding to $A=0.707$ is called cut-off frequency, in this experiment, the high-pass filter and low-pass filter are consisting of RC network, the cut-off frequencies of them are both $1/2\pi RC$. Figure 7-2(c) has response in a frequency band, called band-pass filter, in the figure, f_{C1} is called lower cut-off frequency, f_{C2} is called upper cut-off frequency, passband $BW = f_{C2} - f_{C1}$.

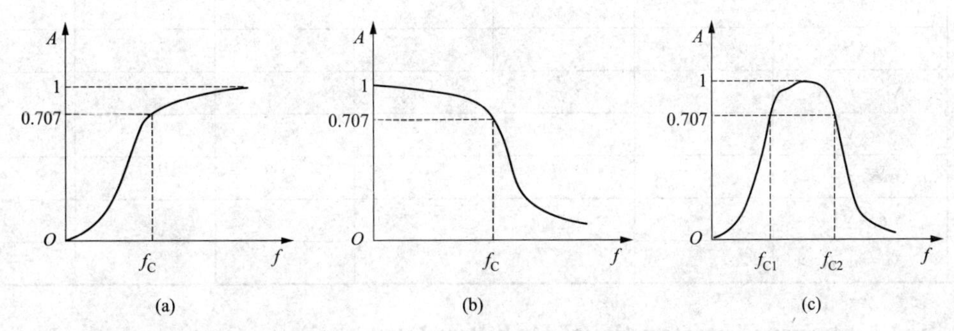

Figure 7-2 The Amplitude Frequency Characteristics of Three Typical Filters
(a) High-Pass Filter; (b) Low-Pass Filter; (c) Band-Pass Filter

● Equipment

Equipment is shown in Table 7-1.

Table 7-1 Equipment

Equipment	Model or Specification	Quantity	Module
Oscilloscope	GDS-1102A-U	1	
Signal Generator	DG1022U	1	
Resistor	300Ω	1	NDG-13
Resistor	1kΩ	1	
Resistor	2kΩ	1	
Inductor	10mH	1	
Capacitor	0.01μF	1	

● Contents

1. Measure the Inductive Reactance Frequency Response of *R*, *L*, *C*

The experiment circuit is shown in Figure 7-1, $r=300\Omega$, $R=1\text{k}\Omega$, $L=10\text{mH}$, $C=0.01\mu\text{F}$. The input voltage u is the sinusoidal wave output of the signal generator, adjust the voltage RMS (Root Mean Square) to 2V according to measure function of the oscilloscope and keep the voltage constant.

Connect R, L, C to r with a wire separately, increase the output frequency of the signal generator from 1kHz to 20kHz gradually, measure u_R, u_L, u_C and u_r with the oscilloscope separately, fill in Table 7 - 2. Calculate the R, X_L and X_C under different frequencies.

Table 7 - 2 Data of the Inductive Reactance Frequency Response of *R*, *L*, *C*

	f(kHz)	1	2	5	10	15	20
R(kΩ)	u_r(V)						
	i_R(mA) $=u_r/r$						
	u_R(V)						
	$R=u_R/i_R$						
X_L(kΩ)	u_r(V)						
	i_L(mA) $=u_r/r$						
	u_L(V)						
	$X_L=u_L/i_L$						
X_C(kΩ)	u_r(V)						
	i_C(mA) $=u_r/r$						
	u_C(V)						
	$X_C=u_C/i_c$						

2. The Frequency Characteristics of High-Pass Filter

The experiment circuit is shown in Figure 7 - 3, $R=2\text{k}\Omega$, $C=0.01\mu\text{F}$. The excitation signal (input voltage) u_i is the sinusoidal wave output of the signal generator. Adjust the voltage RMS to 2V according to measure function of the oscilloscope, and keep the voltage unchanged. Increase the output frequency of the signal generator from 1kHz to 20kHz gradually. Measure the response signal (output voltage) U_R, and fill in Table 7 - 3.

Table 7 - 3 Data of the Frequency Characteristics

f(kHz)	1	3	6	8	10	15	20
U_R(V)							
U_C(V)							
U_O(V)							

3. The Frequency Characteristics of Low-Pass Filter

The experiment circuit and the procedure is the same as step 2, but the response signal (output voltage) is the voltage u_C on the capacitor. Fill in Table 7 - 3.

4. The Frequency Characteristics of Band-Pass Filter

The experiment circuit is shown in Figure 7 - 4, $R=1\text{k}\Omega$, $L=10\text{mH}$, $C=0.1\mu\text{F}$. The procedure is the same as step 2. The response signal (output voltage) is the voltage U_O across the resistor. Fill in Table 7 - 3.

● **Notes**

Pay attention to the correct usage of the oscilloscope.

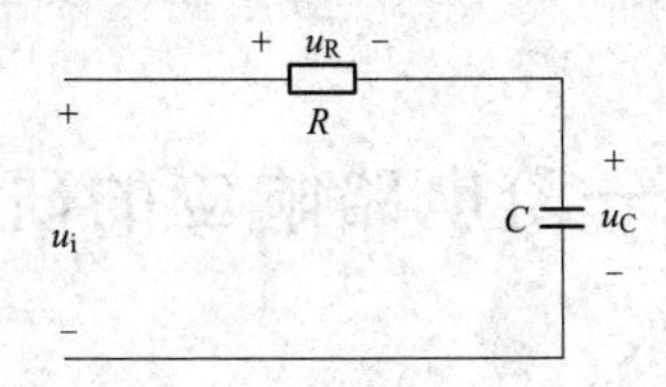

Figure 7 - 3 The Experiment Circuit of the High-Psaa Filter

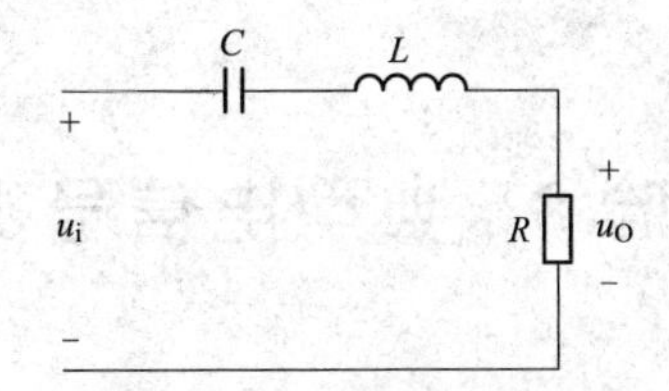

Figure 7 - 4 The Experiment Circuit of the Low-Pass Filter and the Band-Pass Filter

- **Questions**

1. How to measure resistance R, inductance X_L and capacitance X_C? What are the relationships between their values and frequency?

2. What are the frequency characteristics? What are the characteristics of high-pass filter, low-pass filter and band-pass filter? How to measure those characteristics?

- **Experiment Report**

1. Draw the frequency response curves of R, L, C elements on graph paper, summarize the characteristics.

2. Draw the frequency response curves of high-pass filter, low-pass filter and band-pass filter, and summarize them.

实验 8　典型电信号观测与 RC 一阶电路响应的研究

一、实验目的

(1) 了解信号发生器，示波器各旋钮、按键的作用。

(2) 掌握用信号发生器输出特定类型和参数的周期性信号的操作方法，以及用示波器观察电信号波形、定量测出周期性信号波形参数的方法。

(3) 了解 RC 一阶电路的零输入响应、零状态响应和全响应的规律和特点。

(4) 学习一阶电路时间常数的测量方法，了解电路参数对时间常数的影响。

(5) 掌握微分电路和积分电路的基本概念。

二、实验原理

(1) 示波器是一种信号图形观测仪器，可显示电信号波形并测出信号的幅值、周期、脉宽、相位差等波形参数。一台双踪示波器可以同时观察和测量两个信号的波形和参数。结合各调节旋钮与按键，示波器可在不同要求下完成对不同波形的观察和测量，希望在实验中加以摸索和掌握。

(2) RC 一阶电路的零状态响应。RC 一阶电路如图 8-1 所示，初始状态下开关 S 在 1 的位置，$u_C=0$，即处于零状态，当开关 S 合向 2 的位置时，电源通过 R 向电容 C 充电，$u_C(t)$ 称为零状态响应，有

$$u_C = U_S - U_S e^{-\frac{t}{\tau}}$$

零状态响应变化曲线如图 8-2 所示，当 u_C 上升到 $0.632U_S$ 所需要的时间称为时间常数 τ，$\tau=RC$。

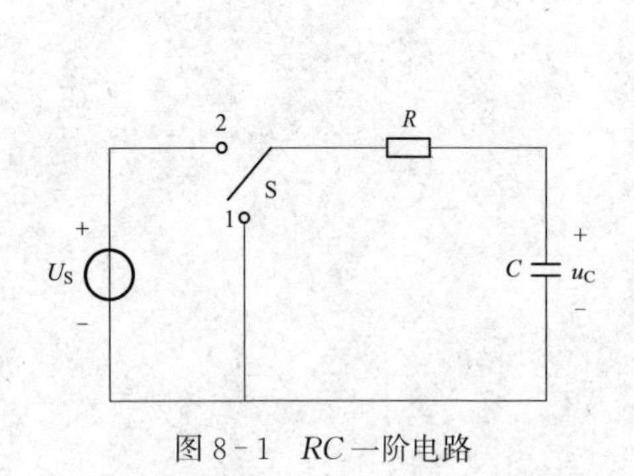

图 8-1　RC 一阶电路

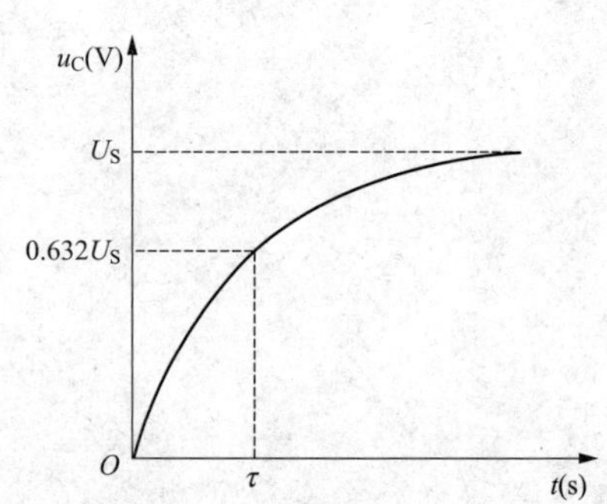

图 8-2　零状态响应变化曲线

(3) RC 一阶电路的零输入响应。在图 8-1 中，开关 S 在 2 的位置电路稳定后，再合向 1 的位置时，电容 C 通过 R 放电，$u_C(t)$ 称为零输入响应，有

$$u_C = U_S e^{-\frac{t}{\tau}}$$

零输入响应变化曲线如图 8-3 所示，当 u_C 下降到 $0.368U_S$ 时所需要的时间为时间常数 τ，$\tau=RC$。

(4) 测量 RC 一阶电路时间常数。动态网络的过渡过程是十分短暂的单次变化过程。要

用普通示波器观察过渡过程和测量有关的参数，就必须使这种单次变化的过程重复出现。为此，利用信号发生器输出的方波来模拟阶跃激励信号，即利用方波输出的上升沿作为零状态响应的正阶跃激励信号；利用方波的下降沿作为零输入响应的负阶跃激励信号。只要选择的方波周期远大于电路的时间常数 τ，那么电路在这样的方波序列脉冲信号的激励下，它的响应就和直流电接通与断开的过渡过程是基本相同的。

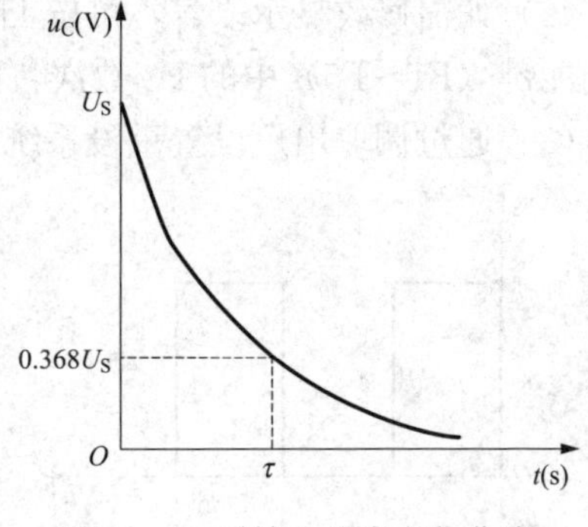

图 8－3　零输入响应变化曲线

实验电路中采用图 8－4 所示的周期性方波 u_S 作为电路的激励信号，方波信号的周期为 T，只要满足 $\frac{T}{2} \geqslant 5\tau$，便可在示波器的屏幕上形成稳定的响应波形。将 R、C 串联与信号发生器的输出端连接，用双踪示波器观察电容电压 u_C，便可观察到稳定的周期性变化曲线，如图 8－5 所示。以零状态响应部分为例，在屏幕上测得电容电压最大值 U_{Cm}，u_C 轴上 $0.632U_{Cm}$ 位置与指数曲线交点对应 t 轴的 x 点，则 0 到 x 点之间为电路的时间常数 τ。

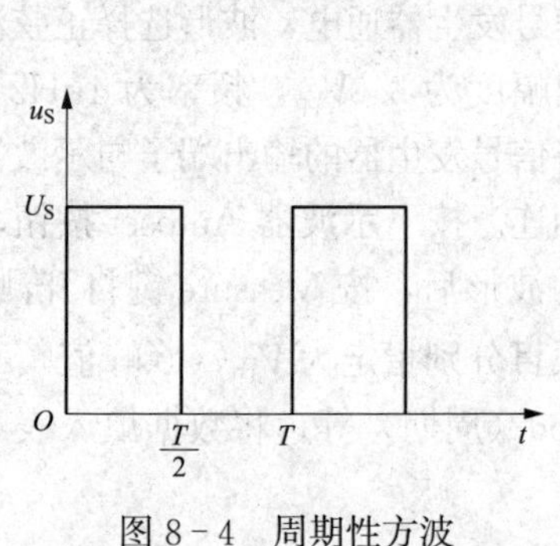

图 8－4　周期性方波

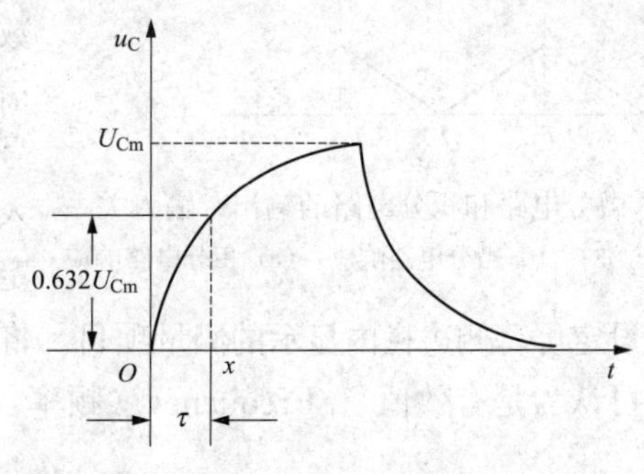

图 8－5　时间常数

（5）微分电路和积分电路。方波信号 u_S 作用在 R、C 串联电路中，当电路时间常数 τ 与方波周期 T 满足 $\tau \ll T/2$ 的条件时，电阻 R 两端电压 u_R 与方波输入信号 u_S 的波形呈近似的微分关系，有 $u_R \approx RC\frac{du_S}{dt}$，该电路称为微分电路。当满足 $\tau \gg T/2$ 的条件时，电容 C 两端电压 u_C 与方波输入信号 u_S 呈近似的积分关系，$u_C \approx \frac{1}{RC}\int u_S dt$，该电路称为积分电路。

微分电路和积分电路的输出、输入关系如图 8－6 所示。

三、实验设备

实验设备见表 8－1。

四、实验内容

1. 数字示波器的设置

以下为全新示波器使用前需进行的简单设置，如实验所用示波器已经由他人使用过，部分步骤可跳过。

（1）连接电源线，打开示波器电源开关，约 10s 后示波器显示正常启动。

（2）如需修改显示语言，可按 Utility 键调出语言菜单，然后按屏幕右侧纵向排列的 5 个功能键（F1～F5）中的 F3 数次，进行切换，将语言切换至中文。

（3）通过调取出厂设置重设系统，按 Save/Recall 键，选择默认设置。

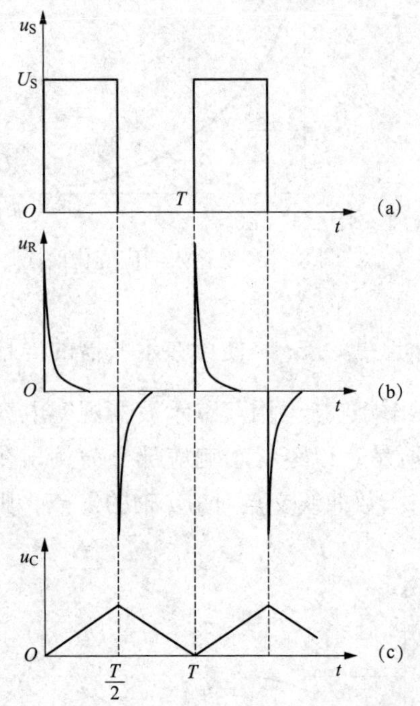

图 8－6 微分电路和积分电路的输出、输入关系
（a）激励方波；（b）微分电路响应；（c）积分电路响应

（4）如使用随机自带探棒，设置探棒自身衰减为×10，用探棒连接通道 1 的输入端和探棒补偿信号（$2V_{p-p}$，1kHz 方波）输出端。

（5）按 Autoset 键，方波显示于屏幕上。

（6）如方波波形有弯曲，使用工具对探棒接口附近螺丝进行调节，至波形平整为止。

示波器的 Help 键可显示各按键功能说明。使用方法为先按下 Help 键，再按其他功能键，屏幕显示对应按键的帮助内容，旋转 Variable 旋钮可滚动帮助内容。再按一次 Help 键退出帮助内容。

2. 正弦波信号的观测

（1）使用示波器自动测量功能测量信号参数。将信号发生器通电，波形选择正弦波，调节信号输出幅度为 3.0V_{pp}，频率为 1kHz。通过信号线缆将信号发生器的输出端子与示波器任一输入通道相连。按下示波器 Autoset 按钮，得到稳定显示的波形后，按 Measure 键打开测量功能，按功能键设定屏幕侧边栏内显示的测量项目，将各个项目分别指定为 V_{pp}（峰峰值）、V_{rms}（均方根值，可认为是有效值）、Frequency（频率）、Period（周期）等，将数据填入表 8－2 中。

表 8－1 实 验 设 备

设备名称	型号与规格	数量	实验模块
双踪示波器	GDS－1102A－U	1	
信号发生器	DG1022U	1	
电阻	100Ω	1	NDG－13
电阻	10kΩ	1	
电阻	100kΩ	1	
电阻	30kΩ 或其他		
电容	0.01μF	1	
电容	0.1μF 或其他		

表 8－2 正弦信号参数（自动测量）

参数	峰峰值	有效值	频率	周期
数值				

（2）使用示波器光标功能测量信号参数。保持示波器输入信号不变，按 Cursor 键，屏

幕上出现光标。按 X→Y 可切换水平和垂直光标，按 Source 可选择信号通道。需要移动水平光标 X1、X2 时，先按 X1、X2 菜单项对应的功能键，然后用 Variable 旋钮移动相应的光标，即可在屏幕侧边栏中读出两光标间的时间差，移动垂直光标 Y1、Y2 的操作方法类似，可读出两光标的电压差。将水平光标移动到波形一个周期的两端，读出周期并计算出频率；将垂直光标移动到波形的最高点和最低点，读出 V_{PP}并计算出有效值，填入表 8-3 中。

表 8-3　正弦信号参数（光标测量）

参数	峰峰值	有效值	频率	周期
数值				

3. 方波信号的观测

将信号发生器波形切换为方波，保持输出幅度和频率不变。按 Measure 键，为菜单栏指定 V_{PP}、频率、周期、+Width（脉宽，即方波高电平持续时间）、Duty Cycle（占空比，即方波高电平占整个周期的比例），将自动测量的数据填入表 8-4 中。

表 8-4　方波信号参数（自动测量）

参数	峰峰值	频率	周期	脉宽	占空比
数值					

利用光标测量方波信号的 V_{PP}、周期、脉宽并计算频率和占空比，填入表 8-5 中。

表 8-5　方波信号参数（光标测量）

参数	峰峰值	频率	周期	脉宽	占空比
数值					

4. 观测一阶电路的充电、放电过程并测量时间常数 τ

实验电路如图 8-7 所示，图中 $R=10\text{k}\Omega$、$C=0.01\mu\text{F}$ 从实验模块上选取和串联。信号源用信号线与 R、C 电路相连接，信号源输出 $V_{PP}=3\text{V}$，$f=1\text{kHz}$ 的方波作为电路的激励。用双踪示波器观察方波信号 u_S和电容 C 上的响应信号 u_C，描绘响应波形。依据实验原理中测量时间常数 τ 的方法，结合示波器光标的使用，测量 τ 值并记录。继续增大 R 或 C 之值，定性的观察对响应的影响。

5. 积分电路和微分电路

(1) 积分电路：实验电路如图 8-7 所示，令 $R=100\text{k}\Omega$、$C=0.01\mu\text{F}$（也可自行指定 R、C 数值，但必须满足积分电路条件），观测并描绘响应 u_C的波形。

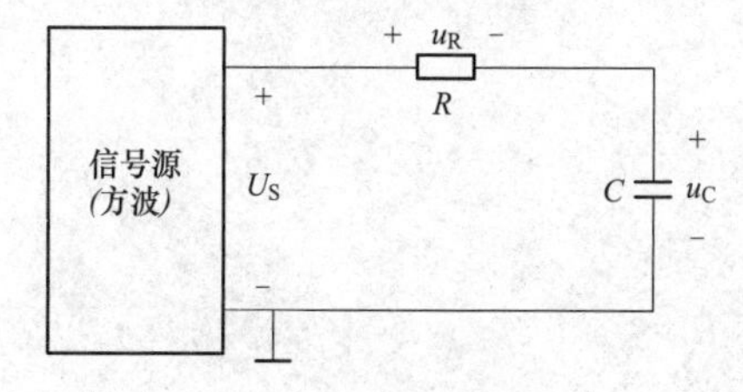

图 8-7　观测一阶电路的充电、放电过程

(2) 微分电路：将图 8-7 中 R、C 元件位置互换，令 $R=100\Omega$、$C=0.01\mu\text{F}$（也可自行指定 R、C 数值，但必须满足微分电路条件），观测电阻 R 上的响应信号 u_R，描绘响应波形。

五、注意事项

(1) 连接信号源的信号线的接地端与连接示波器的信号线接地端要连在一起（称为共地），以防外界干扰而影响测量的准确性。注意微分电路与积分电路共地点的区别。

（2）实验中应充分利用示波器自带的帮助内容，掌握示波器各按钮、旋钮的使用方法。

六、思考题

（1）已知 RC 一阶电路的 $R=10\mathrm{k}\Omega$、$C=6800\mathrm{pF}$，试计算时间常数 τ，并根据 τ 值的物理意义，拟定测量的方案。

（2）在 RC 一阶电路中，当 R、C 的大小变化时，对电路的响应有何影响？

七、实验报告

（1）根据实验内容 4 观测结果，在方格纸上绘出 u_C 与激励信号对应的波形。比较在示波器上测得的 τ 值与计算出的 τ 值，如有误差，分析产生误差的原因。

（2）根据实验内容 5 观测结果，在方格纸上绘出积分电路、微分电路响应信号与激励信号对应的波形。

Experiment 8 Observation of Typical Electric Signals and Responses of First-Order *RC* Circuits Response

● Objectives

1. Understand the functions of knobs and buttons on the signal generator and oscilloscope.

2. Learn how to produce periodic signals of specific types and parameters with a signal generator, and observe the waveforms of electric signals and measure the parameters of periodic signals with an oscilloscope.

3. Understand the laws and characteristics of zero-input response, zero-state response and complete response of *RC* circuit.

4. Learn how to measure the time constant of first-order circuit, and understand the influence of circuit parameters on the time constant.

5. Understand the basic concepts of differential circuits and integral circuits.

● Principles

1. An oscilloscope is an instrument for displaying signal graphics, which can display the waveform of electric signals and measure the amplitude, period, pulse width, phase difference and other parameters of signals. A dual-channel oscilloscope can show the waveforms and the parameters of two signals. With the knobs and buttons, the oscilloscope can display and measure different waveforms according to different requirements.

2. Zero-State Response of an *RC* Circuit

An *RC* circuit is shown in Figure 8 - 1. The switch is in position 1, and the circuit is in zero-state, $u_C = 0$, when the switch moves to position 2, the voltage source U_S begins to charge the capacitor *C*. The zero-state response $u_C(t)$ is

$$u_C = U_S - U_S e^{-\frac{t}{\tau}}$$

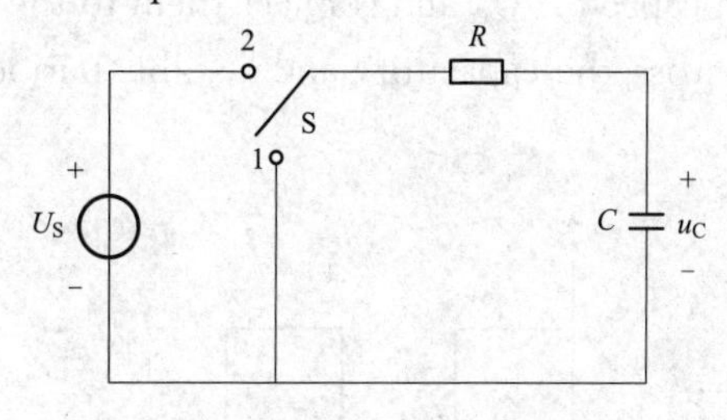

Figure 8 - 1 First-Order *RS* Circuit

The plot of the capacitor voltage u_C is shown in Figure 8 - 2, the time required for u_C to rise to $0.632U_S$ is called the time constant τ, and $\tau = RC$.

3. Zero-Input Response of an *RC* Circuit

The switch in Figure 8 - 1 has been in position 2 for a long time. At $t = 0$, it moves to position 1. The capacitor *C* discharges and the zero-input response $u_C(t)$ is

$$u_C = U_S e^{-\frac{t}{\tau}}$$

The plot of u_C is shown in Figure 8 - 3, and the time constant $\tau = RC$ is the time required for u_C to delay by 36.8 percent of U_S, i. e. $0.368U_S$.

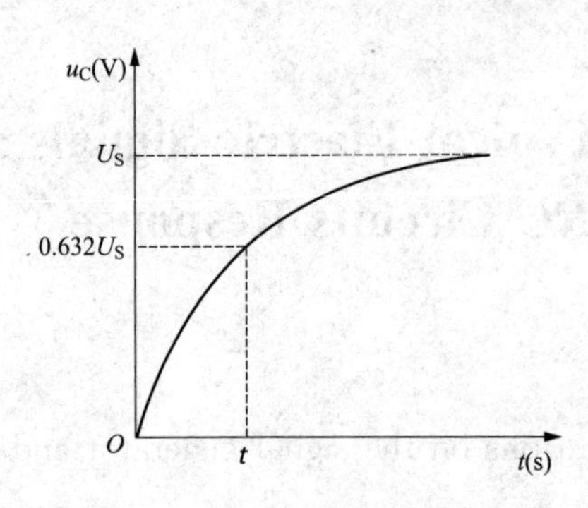

Figure 8 - 2 Zero-State Response

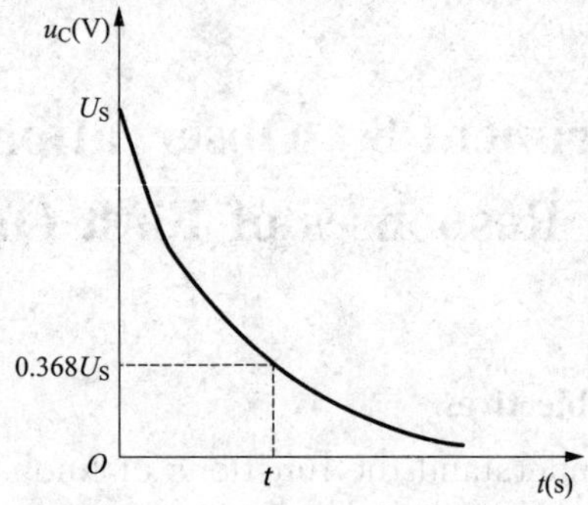

Figure 8 - 3 Zero-Input Response

4. Measuring the Time Constant τ of an RC Circuit

The transition process of a dynamic network is transient. The process must be repeated for observing and measuring them with common oscilloscopes. So, a square wave signal from the signal generator is used for simulating the step function as an excitation. The rising edge of square wave is used as the positive step excitation signal of zero-state response; the drop edge of square wave is used as the negative step excitation signal of zero-input signal. As long as the period of the chosen square wave signal is greater than the time constant τ, the response of the circuit to the square wave signal is basically the same as the transient process when the DC power source in the circuit is turned on and then off.

In the experiment circuit, the periodic square wave u_S shown in Figure 8 - 4 is used as an excitation signal, and the period of the square wave signal is T, as long as $\frac{T}{2} \geqslant 5\tau$ a stable waveform will be formed on the screen of the oscilloscope. Connect the resistor R in series with capacitor C, and connect them to the output of the signal generator. Observe the voltage u_C across the capacitor, and a stable periodic curve as shown in Figure 8 - 5 can be observed.

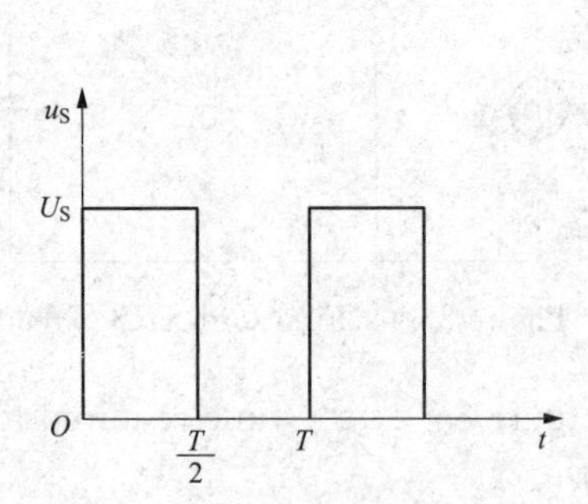

Figure 8 - 4 The Periodic Square Wave

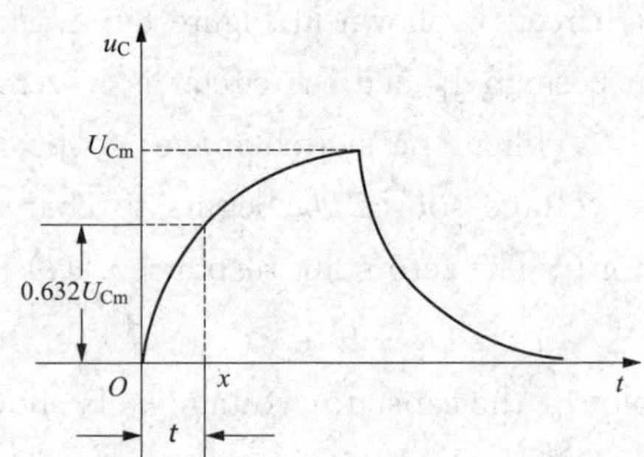

Figure 8 - 5 The Time Constant τ

5. Differential Circuit and Integral Circuit

Consider a series circuit of resistor R and capacitor C excited by a square wave signal. When the time constant τ and the period of the square wave T satisfy the condition $\tau \ll T/2$, the voltage u_R across the resistor R and the input square wave u_S is approximately differential.

$$u_R \approx RC \frac{du_S}{dt}$$

So, the circuit is called the differential circuit.

When the condition $\tau \gg T/2$ is satisfied, the voltage u_C across the capacitor C and the input square wave u_S is approximately integral.

$$u_C \approx \frac{1}{RC}\int u_S \mathrm{d}t$$

So, the circuit is called the integral circuit.

The input square wave, the output waveforms of the differential circuit and integral circuit are shown in Figure 8 - 6.

- **Equipment**

Equipment is shown in Table 8 - 1.

- **Contents**

1. Setting of Oscilloscope

The following contents are the simple settings before the use of a new oscilloscope. If the oscilloscope used in the experiment has already been used in other experiments, some of the following steps can be skipped.

(1) Connect the power cord and press the power switch, the display will become active in approximately 10 seconds.

(2) If the displayed language is not the language you want, press Utility to call up the language menu and press the third function key (F3) among the five function keys (F1～F5) vertically arranged on the right side of the screen to change language.

(3) Reset the system by recalling the factor settings. Press the Save/Recall button, then Default Setup.

(4) If the oscilloscope's probe is used, set the probe attenuation to ×10, connect the probe between the CH1 input terminal and probe compensation output terminal ($2V_{p-p}$, 1kHz square wave).

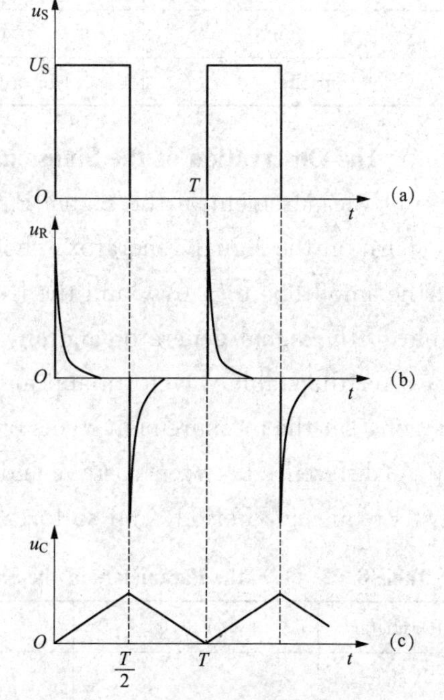

Figure 8 - 6 The Relations between the Input Square Waue and the Output of the Differential Circuit and the Integral Cicuit

(a) The Excitation Square Wave; (b) The Response of the Differential Circuit; (c) The Response of the Integral Circuit

(5) Press Autoset button, a square waveform will appear on the screen.

(6) Turn the adjustment point on the probe to make the square waveform edge even.

The Help key on the oscilloscope shows the descriptions of all keys. Press Help key first, and then press another function key, the help content of the key will be displayed on the screen. Turn the Variable knob, the help content scrolls on the screen. Press Help key again to exit the help content.

Table 8 - 1 Equipment

Equipment	Model or Specification	Quantity	Module
Oscilloscope	GDS - 1102A - U	1	
Signal Generator	DG1022U	1	
Resistor	100Ω	1	NDG - 13
Resistor	10kΩ	1	
Resistor	100kΩ	1	
Resistor	30kΩ or Other Resistors		
Capacitor	0. 01μF	1	
Capacitor	0. 1μF or Other Capacitors		

2. The Observation of the Sinusoidal Wave Signal

(1) Measurement of the Signal Parameters With the Automatic Measurement Function

Turn on the signal generator, choose the sinusoidal wave as the output waveform, adjust the amplitude to 3. $0V_{pp}$ and the frequency to 1kHz. Connect the signal wire between the output of the signal generator and any input terminal of the oscilloscope. Press the Autoset key, after the stable waveform appears, press the Measure key to activate the measurement function, set the measurement items on the side menus with the function keys, set the items as V_{pp} (difference between positive and negative peak voltage), V_{rms} (root mean square voltage), Frequency, Period, and so forth, fill in Table 8 - 2.

Table 8 - 2 the Parameters of the Sinusoidal Signal (Automatic Measurement)

Parameter	V_{PP}	V_{rms}	Frequency	Period
Value				

(2) Measurement of The Signal Parameters With the Cursor Function

Keep the input to the oscilloscope unchanged. Press the Cursor key, and the cursors appear in the display. Press X→Y to switch the horizontal and vertical cursor, and press Source to select the source channel. To move the horizontal cursors X1 and X2, press the function key beside the menu item, then move the cursor with the Variable knob, the time difference between X1 and X2 can be read from the side menu. The voltage difference between the vertical cursors Y1 and Y2 can be read by similar method. Move the horizontal cursors to both ends of a period of the waveform, read the period and calculate the frequency; move the vertical cursors to the highest and the lowest point on the waveform, read V_{PP} and calculate V_{rms}, and fill in Table 8 - 3.

Table 8 - 3 The Parameters of the Sinusoidal Signal (Cursor Measurement)

Parameter	V_{PP}	V_{rms}	Frequency	Period
Value				

3. Observation of a Square Wave Signal

Switch the output waveform to square wave, keep V_{PP} and V_{rms} unchanged. Press the Measure key, and set the menu items to V_{PP}, Frequency, Period, +Width (pulse width, positive pulse width), Duty Cycle (ratio of positive pulse of whole cycle), and fill in Table 8-4 with the automatically measured data.

Table 8-4 the Parameters of the Square Wave (Automatic Measurement)

Parameter	V_{PP}	Frequency	Period	Pulse Width	Duty Cycle
Value					

Measure V_{PP}, period, pulse width with cursors, calculate frequency and duty cycle, and fill in Table 8-5.

Table 8-5 The Parameters of the Square Wave (Cursor Measurement)

Parameter	V_{PP}	Frequency	Period	Pulse Width	Duty Cycle
Value					

4. Observation of the Charging and Discharging Processes and Measure the Time Constant τ

The experiment circuit is shown in Figure 8-7. Choose $R = 10\ \text{k}\Omega$ and $C = 0.01\mu\text{F}$ on the experiment module, and connect the two elements in series. Connect the signal generator and the *RC* series circuit with signal wire, set the output of the signal generator to $V_{PP} = 3\text{V}$ and $f = 1\text{kHz}$ square wave, as the excitation of the circuit. Observe the square wave signal u_S and the response signal u_C across the capacitor C, and draw the waveform of the response. According to the measurement of the time constant τ introduced in the previous experiment principle, using cursors of the oscilloscope, measure and write down the value of τ. Increasing the resistance or capacitance, observe the influence of the time constant τ on the response qualitatively.

5. Integral Circuit and Differential Circuit

(1) Integral circuit: the experiment circuit is shown in Figure 8-7 where $R = 100\text{k}\Omega$ and $C = 0.01\mu\text{F}$. The values of resistance and capacitance can be designated otherwise if satisfying the condition of integral circuit. Observe and draw the waveform of response u_C.

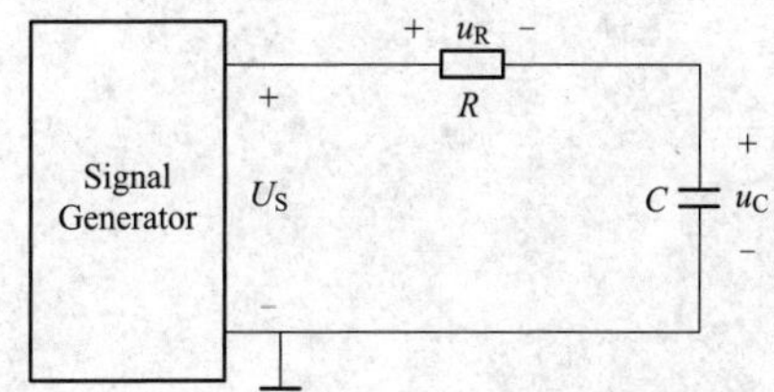

Figure 8-7 Observation of the Charging and Discharging Processes in a First-Order *RC* Circuit

(2) Differential circuit: interchange R and C in Figure 8-7 where $R = 100\Omega$ and $C = 0.01\mu\text{F}$. The values of resistance and capacitance can be designated otherwise if satisfying the condition of differential circuit. Observe the response u_R across the resistor R and draw the waveform.

● Notes

1. The ground terminal of the signal wire connected to the signal generator and the ground terminal of the signal wire connected to the oscilloscope must be connected to each other to form common-ground in order to avoid the external interference affecting the accura-

cy of measurement. Notice that the difference between the common-ground of differential circuit and that of integral circuit.

2. During the experiment process, make full use of the help contents of the oscilloscope's own, and know how to use the knobs and buttons on the oscilloscope.

- **Questions**

1. In a first-order RC circuit, with $R = 10\text{k}\Omega$ and $C = 6800\text{pF}$, calculate the time constant τ, and make a measurement plan of τ according to the physical significance of the time constant.

2. In a first-order RC circuit, when the resistance and capacitance change, how does the response in the circuit change?

- **Experiment Report**

1. Draw the waveform of u_C corresponding to the excitation signal on the graph paper according to the result of step 4. Compare the time constant τ measured on the oscilloscope and the calculated τ if there is error between them, analyze the cause of error.

2. Draw the waveforms of the response signals of the differential circuit and integral circuit corresponding to the excitation signal on the graph paper according to the result of step 5.

实验9　二阶动态电路响应的研究

一、实验目的

(1) 研究 *RLC* 二阶电路的零输入响应、零状态响应的规律和特点，了解电路参数对响应的影响。

(2) 学习二阶电路衰减系数、振荡频率的测量方法，了解电路参数对它们的影响。

(3) 观察、分析二阶电路响应的三种变化曲线及其特点，加深对二阶电路响应的认识与理解。

二、实验原理

1. 零状态响应

在图 9-1 所示 *RLC* 电路中，$u_C(0)=0$，在 $t=0$ 时开关 S 闭合，电压方程为

$$LC\frac{d^2u_C}{dt}+RC\frac{du_C}{dt}+u_C=U$$

这是一个二阶常系数非齐次微分方程，该电路称为二阶电路，电源电压 U 为激励信号，电容两端电压 u_C 为响应信号。根据微分方程理论，u_C 包含暂态分量 u''_C 和稳态分量 u'_C 两个分量，即 $u_C=u''_C+u'_C$，具体解与电路参数 R、L、C 有关。

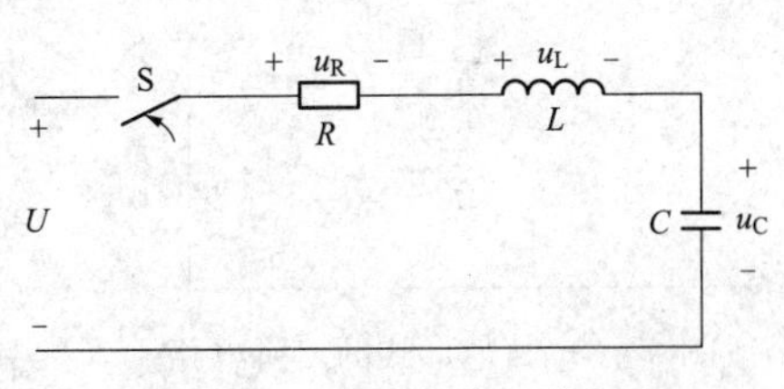

图 9-1　*RLC* 串联二阶电路

当满足 $R<2\sqrt{\frac{L}{C}}$ 时，$u_C(t)=u''_C+u'_C=Ae^{-\delta t}\sin(\omega t+\varphi)+U$

其中，衰减系数 $\delta=\frac{R}{2L}$，衰减时间常数 $\tau=\frac{1}{\delta}=\frac{2L}{R}$，振荡频率 $\omega_d=\sqrt{\frac{1}{LC}-\left(\frac{R}{2L}\right)^2}$，振荡周期 $T=\frac{1}{f}=\frac{2\pi}{\omega_d}$。

变化曲线如图 9-2(a) 所示，u_C 的变化处在衰减振荡状态，由于电阻 R 比较小，又称为欠阻尼状态。

当满足 $R>2\sqrt{\frac{L}{C}}$ 时，u_C 的变化处在过阻尼状态，由于电阻 R 比较大，电路中的能量被电阻很快消耗掉，u_C 无法振荡，变化曲线如图 9-2(b) 所示。

当满足 $R=2\sqrt{\frac{L}{C}}$ 时，u_C 的变化处在临界阻尼状态，变化曲线如图 9-2(c) 所示。

2. 零输入响应

在图 9-3 电路中，开关 S 与 1 端闭合，电路处于稳定状态，$u_C(0)=U$，在 $t=0$ 时开关 S 与 2 闭合，输入激励为零，电压方程为

$$LC\frac{d^2u_C}{dt}+RC\frac{du_C}{dt}+u_C=0$$

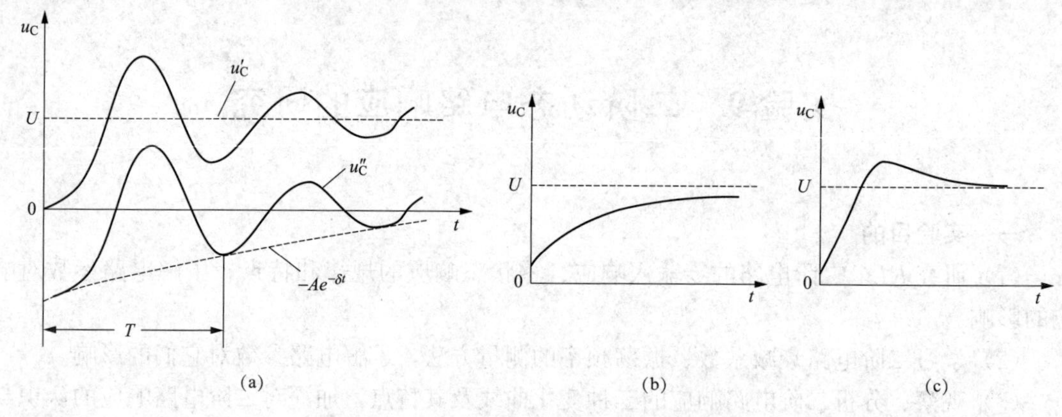

图 9-2 u_C随时间变化的三种状态

(a) 欠阻尼状态；(b) 过阻尼状态；(c) 临界阻尼状态

这是一个二阶常系数齐次微分方程，根据微分方程理论，u_C只包含暂态分量 u''_C，稳态分量 u'_C 为零。和零状态响应一样，根据 R 与 $2\sqrt{\dfrac{L}{C}}$ 的大小关系，u_C 的变化规律分为衰减振荡（欠阻尼）、过阻尼和临界阻尼三种状态，它们的变化曲线与图 9-2 中的暂态分量 u''_C 类似，衰减系数、衰减时间常数、振荡频率与零状态响应完全一样。

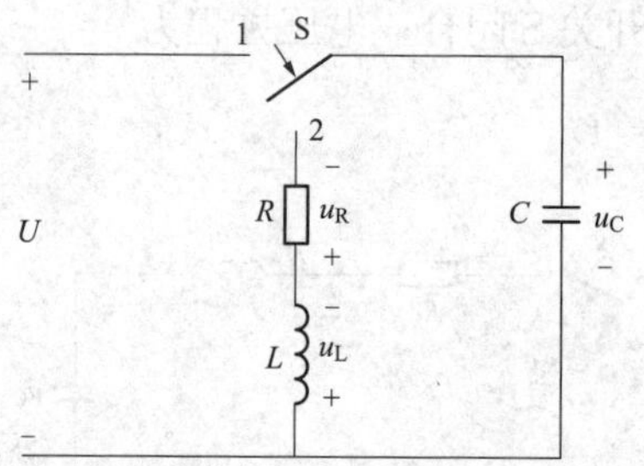

图 9-3 其他形式的二阶电路

本实验对 R、L、C 并联电路进行研究，激励采用方波脉冲，二阶电路在方波正、负阶跃信号的激励下，可获得零状态与零输入响应，响应的规律与 R、L、C 串联电路相同。测量 u_C 衰减振荡的参数，如图 9-2(a) 所示，用示波器测出振荡周期 T，便可计算出振荡频率 ω_d；按照衰减轨迹曲线，测量$-0.368A$ 对应的时间 τ，便可计算出衰减系数 δ。

三、实验设备

实验设备见表 9-1。

表 9-1　　实　验　设　备

设备名称	型号与规格	数量	实验模块
双踪示波器	GDS-1102A-U	1	
信号发生器	DG1022U	1	
电阻	10kΩ	1	NDG-13
电感	15mH	1	
电容	0.01μF	1	
电位器	10kΩ	1	

四、实验内容

实验电路如图 9-4 所示，其中：R_1=10kΩ，L=15mH，C=0.01μF，R_2为 10kΩ 电位器，信号源的输出为最大值 U_m=2V，频率 f=1kHz 的方波脉冲，通过插头接至实验电路的激励端，同时用示波探头将激励端和响应输出端接至双踪示波器两个输入通道。

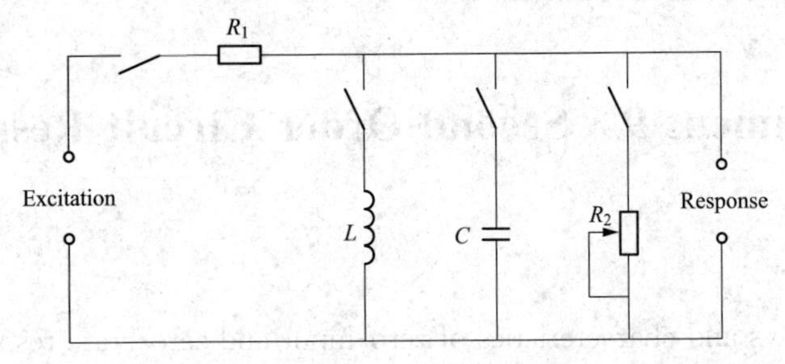

图 9-4　*RLC* 并联二阶电路

（1）调节电位器 R_2，观察二阶电路的零输入响应和零状态响应由过阻尼过渡到临界阻尼，最后过渡到欠阻尼的变化过渡过程，分别定性地描绘响应的典型变化波形。

（2）调节 R_2 使示波器荧光屏上呈现稳定的欠阻尼响应波形，定量测定此时电路的衰减系数 δ 和振荡频率 ω_d，并记入表 9-2 中。

（3）按表 9-2 中的数据改变电路参数，重复步骤 2 的测量，仔细观察改变电路参数时 δ 和 ω_d 的变化趋势，并将数据记入表 9-2 中。

表 9-2　　阶电路暂态过程实验数据

电路参数 实验序列	元件参数				测量值	
	R_1（kΩ）	R_2	L（mH）	C	δ	ω
1	10	调至欠阻尼状态	15	0.01μF		
2	10		15	1000 pF		
3	10		15	3300 pF		
4	30		15	0.01μF		

五、注意事项

调节电位器 R_2 时，要细心、缓慢，临界阻尼状态要找准。

六、思考题

（1）根据二阶电路实验电路元件的参数，计算出处于临界阻尼状态的 R_2 之值。

（2）在示波器屏幕上，如何测得二阶电路零状态响应和零输入响应欠阻尼状态的衰减系数 δ 和振荡频率 ω_d？

七、实验报告

（1）根据观测结果，在方格纸上描绘二阶电路过阻尼、临界阻尼和欠阻尼的响应波形。

（2）测算欠阻尼振荡曲线上的衰减系数 δ、衰减时间常数 τ、振荡周期 T 和振荡频率 ω_d。

（3）归纳、总结电路元件参数的改变，对响应变化趋势的影响。

Experiment 9 Second-Order Circuit Responses

● Objectives

1. Study the laws and characteristics of zero-input and zero-state response of second-order *RLC* circuit. Know the influence of circuit parameters on the responses.

2. Know how to measure attenuation coefficient and oscillation frequency of a second-order circuit, understand the influence of circuit parameters on them.

3. Observe and analyze the three curves of second-order circuit responses and their characteristics, deepen the understanding of the second-order circuit responses.

● Principles

1. Zero-State Response

In the series *RLC* circuit shown in Figure 9 - 1, $u_C(0)=0$, the switch S is closed at $t=0$. Applying KVL for $t>0$

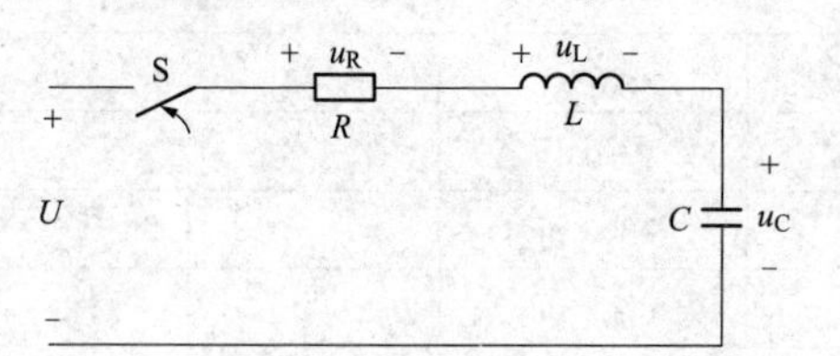

Figure 9 - 1 Series *RLC* Second-Order Circuit

$$L\frac{\mathrm{d}i}{\mathrm{d}t}+Ri+u_C=U \qquad ①$$

But

$$i=C\frac{\mathrm{d}u_C}{\mathrm{d}t}$$

Substituting for i in equation ① and rearranging terms, we get a second-order nonhomogeneous differential equation.

$$LC\frac{\mathrm{d}^2u_C}{\mathrm{d}t}+RC\frac{\mathrm{d}u_C}{\mathrm{d}t}+u_C=U \qquad ②$$

The solution of equation ② has two components, the transient response u''_C and the steady response u'_C, that is $u_C=u''_C+u'_C$, the solution to equation ② is related to circuit parameters R, L, C.

(1) Underdamped Case

For $R<2\sqrt{\frac{L}{C}}$, the zero-state response is

$$u_C(t)=u''_C+u'_C=A\mathrm{e}^{-\delta t}\sin(\omega t+\varphi)+U$$

Where the attenuation coefficient $\delta=\frac{R}{2L}$, the attenuation time constant $\tau=\frac{1}{\delta}=\frac{2L}{R}$, the oscillation frequency $\omega=\sqrt{\frac{1}{LC}-\left(\frac{R}{2L}\right)^2}$, and the oscillation period $T=\frac{1}{f}=\frac{2\pi}{\omega}$.

The curve is shown in Figure 9 - 2(a), and u_C is in underdamped oscillation state, the resistance R is small.

(2) Overdamped Case

For $R > 2\sqrt{\frac{L}{C}}$, the zero-state response is

$$u_C = u''_C + u'_C = A_1 e^{p_1 t} + A_2 e^{p_2 t} + U$$

The curve is shown in Figure 9 - 2(b) and u_C is in overdamped state.

(3) Critically Damped Case

For $R = 2\sqrt{\frac{L}{C}}$ is satisfied, the zero-state response is

$$u_C = u''_C + u'_C = (A_1 + A_2 t)e^{-\delta t} + U$$

The curve is shown in Figure 9 - 2(c) and u_C is in critically damped state.

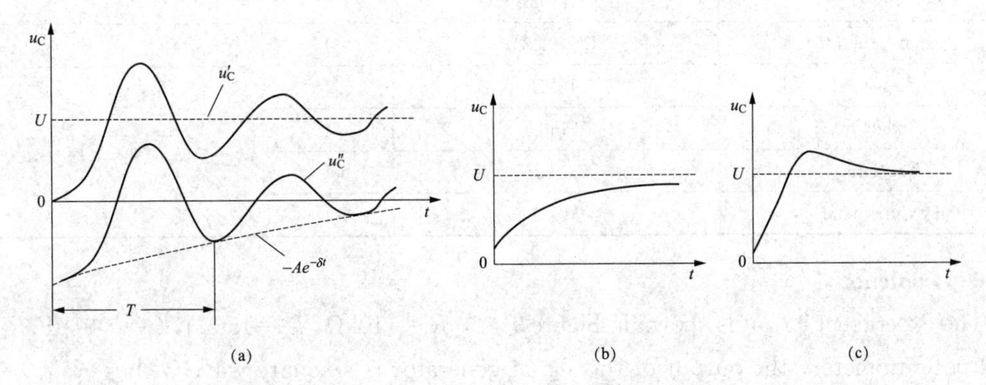

Figure 9 - 2 The Three Cases of u_C Varying With Time

(a) Underdamped Case; (b) Overdamped Case; (c) Critically Damped Case

2. Zero-Input Response

In the circuit of Figure 9 - 3, the circuit is in steady state and $u_C(0) = U$. The switch S moves to terminal 2 at the time $t = 0$, the input excitation is zero, and the voltage equation is

$$LC\frac{d^2 u_C}{dt} + RC\frac{du_C}{dt} + u_C = 0$$

This is a second-order homogeneous differential equation with constant coefficients. According to the differential equation theory, u_C consists only of the transient component u''_C and the steady component u'_C is zero. As in the zero-state response, u_C is divided into underdamped, overdamped and critically damped state according to the relation between R and $2\sqrt{\frac{L}{C}}$, and their curves are similar to those of the transient component u''_C in Figure 9 - 2. The attenuation coefficient, the attenuation time constant, and the oscillation frequency are equal to those of zero-state response, respectively.

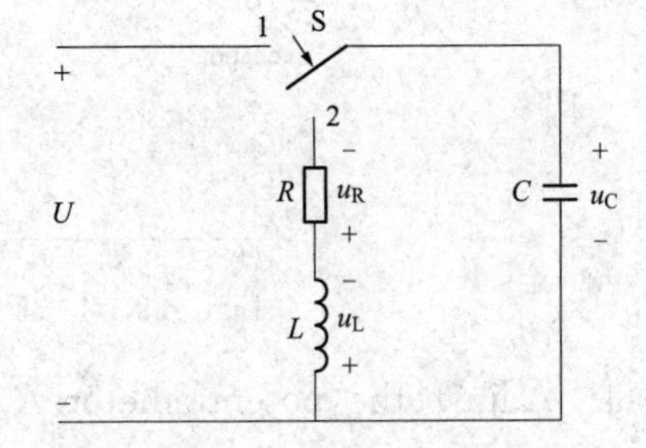

Figure 9 - 3 Another Form of Second-Order Circuit

The parallel RLC circuit is studied in this experiment, and the zero-state and zero-input responses of a second-order circuit to the positive and negative step signals of square wave can be obtained, the laws of response is the same as that of RLC circuit. The plot of u_C, which is exponentially damped and oscillatory, is shown in Figure 9 - 2(a). The damping frequency ω_d can be calculated by the damping period T measured with an oscilloscope. The damping coefficient δ can be calculated by τ measured at $-0.368A$ on the plot in Figure 9 - 2(a).

● Equipment

Equipment is shown in Table 9 - 1.

Table 9 - 1 Equipment

Equipment	Model or Specification	Quantity	Module
Oscilloscope	GDS - 1102A - U	1	
Signal Generator	DG1022U	1	
Resistor	10kΩ	1	NDG - 13
Inductor	15mH	1	
Capacitor	0.01μF	1	
Potentiometer	10kΩ	1	

● Contents

The experiment circuit is shown in Figure 9 - 4, $R_1 = 10\text{k}\Omega$, $L = 15\text{mH}$, $C = 0.01\mu\text{F}$. R_2 is a 10kΩ potentiometer, the output of the signal generator is a square wave with $U_m(V_{PP}) = 2\text{V}$ and $f = 1\text{kHz}$. Connect the output to the excitation port of the experiment circuit, and connect the excitation and the response to the oscilloscope with probes.

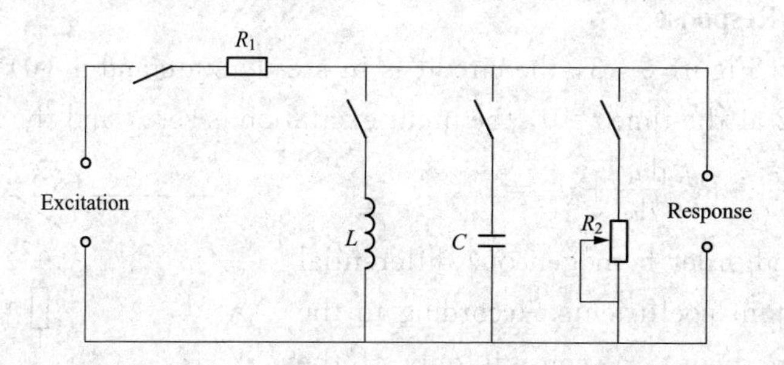

Figure 9 - 4 The Parallel RLC Second-Order Circuit

1. Adjust the potentiometer R_2, observe the transient processes of the zero-input response and zero-state response of the second-order circuit changing from overdamped to critically damped and then to underdamped case, and draw the typical variation waveforms of the responses separately and qualitatively.

2. Adjust R_2 until a steady waveform of underdamped response appears on the screen of the oscilloscope, measure the attenuation coefficient δ and the oscillation frequency ω of the circuit at this point quantitatively, and fill in Table 9 - 1.

3. Change the parameters of the circuit according to Table 9 - 1, repeat step 2, carefully observe the variation trend of δ and ω_d when the parameters is changed, and fill in Table 9 - 2.

Table 9 - 2 Experiment Data of the Transient Process of the Second-Order Circuit

Parameters No.	Parameters of the Elements				Measured Values	
	R_1 (kΩ)	R_2	L (mH)	C	δ	ω_d
1	10	InUnderdamped State	15	0.01μF		
2	10		15	1000pF		
3	10		15	3300pF		
4	30		15	0.01μF		

- **Notes**

Adjust the potentiometer R_2 carefully and slowly to find the accurate critically damped case.

- **Questions**

1. Calculate the value of R_2 in the critically damped state according to the parameters of the elements in the second-order experiment circuit.

2. How to measure the attenuation coefficient δ and the oscillation frequency ω of the zero-state response and the zero-input response of the second-order circuit in the under damping state in the display of the oscilloscope?

- **Experiment Report**

1. Draw the response waveform of the overdamped, critically damped and underdamped case on the graph paper, respectively.

2. Measure the attenuation coefficient δ, the attenuation time constant τ, the oscillation period T and the oscillation frequency ω.

3. Summarize the influence of the parameters of the elements on the changing trend of the response.

实验 10　*RC* 选频网络特性测试

一、实验目的

(1) 研究 *RC* 串-并联电路及 *RC* 双 T 电路的频率特性。

(2) 学会使用示波器测定 *RC* 网络的幅频特性和相频特性。

(3) 熟悉文氏电桥电路的结构特点及选频特性。

二、实验原理

图 10-1 所示 *RC* 串-并联电路的传递函数

$$N(\mathrm{j}\omega)=\frac{\dot{U}_\mathrm{o}}{\dot{U}_\mathrm{i}}=\frac{1}{3+\mathrm{j}\left(\omega RC-\dfrac{1}{\omega RC}\right)}$$

其中幅频特性为

$$A(\omega)=\frac{U_\mathrm{o}}{U_\mathrm{i}}=\frac{1}{\sqrt{3^2+\left(\omega RC-\dfrac{1}{\omega RC}\right)^2}}$$

相频特性为

$$\varphi(\omega)=\varphi_\mathrm{o}-\varphi_i=-\arctan\frac{\omega RC-\dfrac{1}{\omega RC}}{3}$$

幅频特性和相频特性曲线如图 10-2 所示，幅频特性呈带通特性。

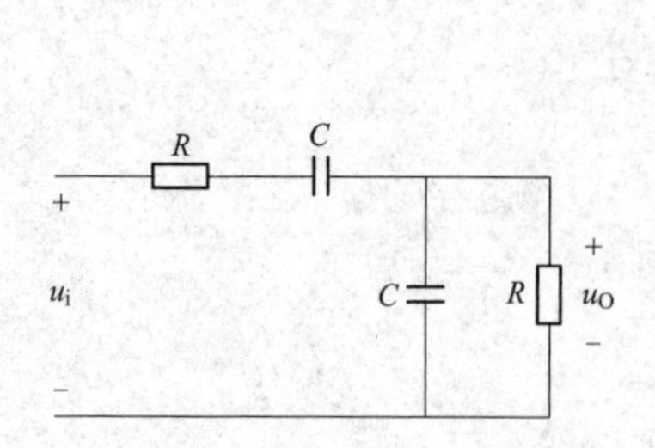

图 10-1　*RC* 串-并联电路

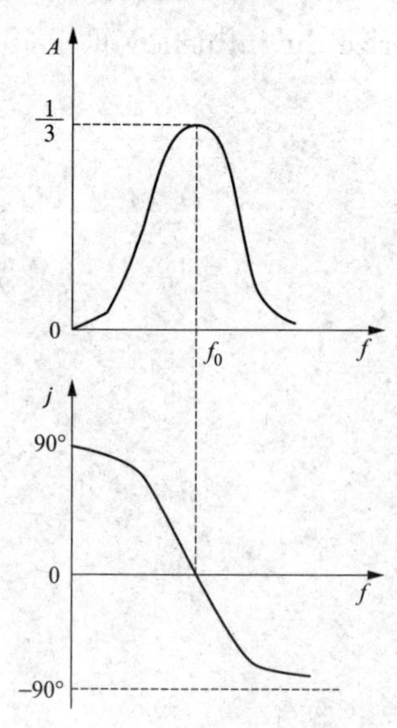

图 10-2　*RC* 串-并联电路的幅频特性和相频特性曲线

当角频率 $\omega=\dfrac{1}{RC}$ 时，$A(\omega)=\dfrac{1}{3}$，$\varphi(\omega)=0°$，u_O与 u_i同相，即电路发生谐振，谐振频

率 $f_0=\dfrac{1}{2\pi RC}$。也就是说，当信号频率为 f_0时，RC 串-并联电路的输出电压 u_O与输入电压 u_i同相，其大小是输入电压的 1/3，这一特性称为 RC 串-并联电路的选频特性，该电路又称为文氏电桥。

测量幅频特性：保持信号源输出电压（即 RC 网络输入电压）U_i恒定，改变频率 f，用示波器监视 U_i，并测量对应的 RC 网络输出电压 U_O，计算出它们的比值 $A=U_O/U_i$。以 f 为横轴、U_O为纵轴，逐点描绘出幅频特性曲线。

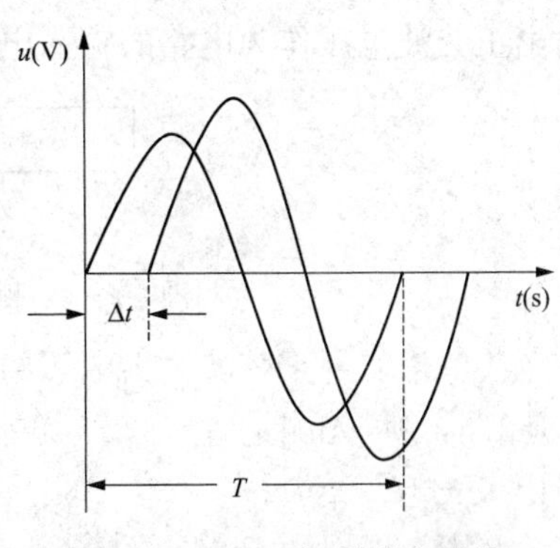

图 10-3　RC 串-并联电路的输入与输出波形

测量相频特性：保持信号源输出电压（即 RC 网络输入电压）U_i恒定，改变频率 f，用示波器监视 U_i，用双踪示波器观察 u_O与 u_i波形。如图 10-3 所示为输入与输出波形的一种情况，若两个波形的延时为 Δt，周期为 T，则它们的相位差 $\varphi=\dfrac{\Delta t}{T}\times 360°$。以 f 为横轴、φ 为纵轴，逐点描绘出相频特性曲线。

用同样方法可以测量 RC 双 T 电路的幅频特性，RC 双 T 电路如图 10-4 所示，其幅频特性具有带阻特性，如图 10-5 所示。

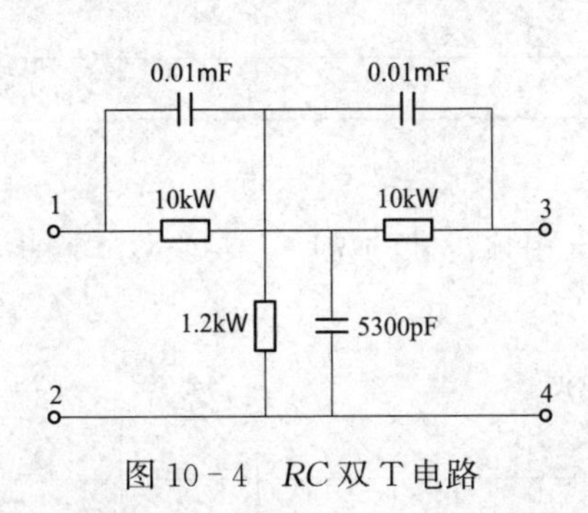

图 10-4　RC 双 T 电路

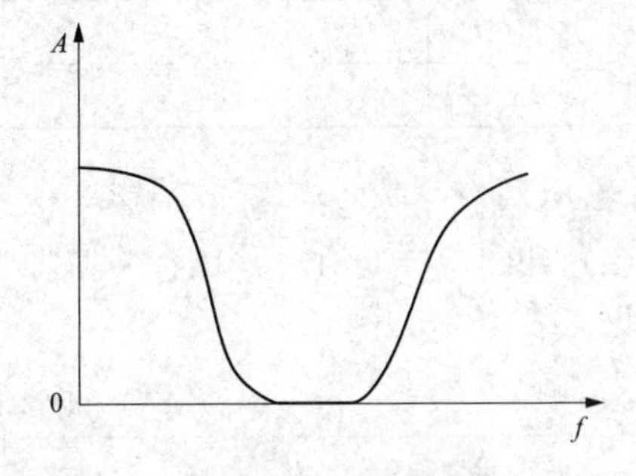

图 10-5　RC 双 T 电路的幅频特性曲线

三、实验设备

实验设备见表 10-1。

表 10-1　**实 验 设 备**

设备名称	型号与规格	数量	实验模块
双踪示波器	GDS-1102A-U	1	
信号发生器	DG1022U	1	
实验电路	选频电路	1	NDG-12
实验电路	双 T 网络	1	
电阻	200Ω	1	NDG-13
电阻	2kΩ	1	
电容	0.22μF	1	
电容	2.2μF	1	

四、实验内容

1. 测量 RC 串-并联电路的幅频特性

实验电路结构如图 10－6 所示，图中箭头代表信号源/示波器探头线。RC 网络按图 10－1 连接或选取 NDG－12 上的选频电路。RC 网络的参数选择为 $R=200\Omega$，$C=2.2\mu F$。信号源输出正弦波电压作为电路的输入电压 u_i，调节信号源输出电压幅值，使 $U_i=2V$。

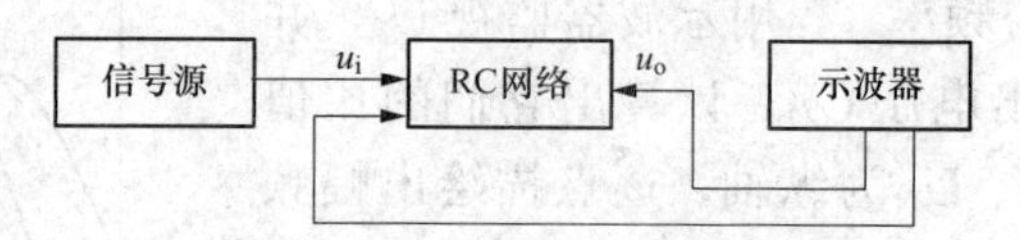

图 10－6　测量 RC 串-并联电路的频率特性实验框图

改变信号源正弦波输出电压的频率 f，并保持 $U_i=2V$ 不变（用示波器监视），测量输出电压 U_O，（可先测量 $A=1/3$ 时的频率 f_0，然后再在 f_0 左右选几个频率点，测量 U_O），将数据记入表 10－2 中。

在图 10－6 的 RC 网络中，选取另一组参数 $R=2k\Omega$，$C=0.22\mu F$，重复上述测量，将数据记入表 10－2 中。

表 10－2　　幅频特性数据

$R=2k\Omega$，$C=0.22\mu F$	f (Hz)									
	U_O (V)									
$R=200\Omega$ $C=2.2\mu F$	f (Hz)									
	U_O (V)									

2. 测量 RC 串-并联电路的相频特性

实验电路如图 10－6 所示，按实验原理中测量相频特性的说明，实验步骤同步骤 1，将实验数据记入表 10－3 中。

表 10－3　　相频特性数据

$R=200\Omega$，$C=2.2\mu F$	f (Hz)									
	T (ms)									
	Δt (ms)									
	φ									
$R=2k\Omega$ $R=1k\Omega$ $C=0.1\mu F$	f (Hz)									
	T (ms)									
	Δt (ms)									
	φ									

3. 测定 RC 双 T 电路的幅频特性

实验电路如图 10－6 所示，其中 RC 网络按图 10－4 连接或选取 NDG－12 上的双 T 网络。实验步骤同步骤 1，将实验数据记入自拟的数据表格中。

五、注意事项

由于信号源内阻的影响，注意在调节输出电压频率时，应同时调节输出电压大小，使实验电路的输入电压保持不变。

六、思考题

（1）根据电路参数，估算 *RC* 串-并联电路两组参数时的谐振频率。

（2）推导 *RC* 串-并联电路的幅频、相频特性的数学表达式。

（3）什么是 *RC* 串-并联电路的选频特性？当频率等于谐振频率时，电路的输出、输入有何关系？

（4）试定性分析 *RC* 双 T 电路的幅频特性。

七、实验报告

（1）根据表 10-2 和表 10-3 实验数据，绘制 *RC* 串-并联电路的两组幅频特性和相频特性曲线，找出谐振频率和幅频特性的最大值，并与理论计算值比较。

（2）设计一个谐振频率为 1kHz 文氏电桥电路，说明它的选频特性。

（3）根据实验 3 的实验数据绘制 *RC* 双 T 电路的幅频特性曲线，并说明幅频特性的特点。

Experiment 10 Testing Selectivity Characteristics of *RC* Network

- **Objectives**

1. Study the frequency response of series *RC* circuit, parallel *RC* circuit and *RC* double-T circuit.

2. Learn how to measure the amplitude response and the phase response of *RC* network with an oscilloscope.

3. Study the structure characteristic and the frequency selectivity of Wien bridge.

- **Principles**

The transfer function of series-parallel *RC* circuit shown in Figure 10 - 1 is

$$N(\mathrm{j}\omega)=\frac{\dot{U}_{\mathrm{o}}}{\dot{U}_{\mathrm{i}}}=\frac{1}{3+\mathrm{j}\left(\omega RC-\dfrac{1}{\omega RC}\right)}$$

We obtain the magnitude and phase of $N(\mathrm{j}\omega)$ as

$$A(\omega)=\frac{U_{\mathrm{o}}}{U_{\mathrm{i}}}=\frac{1}{\sqrt{3^2+\left(\omega RC-\dfrac{1}{\omega RC}\right)^2}}$$

$$\varphi(\omega)=\varphi_{\mathrm{o}}-\varphi_i=-\arctan\frac{\omega RC-\dfrac{1}{\omega RC}}{3}$$

The amplitude response and the phase response are shown in Figure 10 - 2 respectively. It is clear from the amplitude response $A(\omega)$ that the circuit in Figure 10 - 1 is band-pass filter.

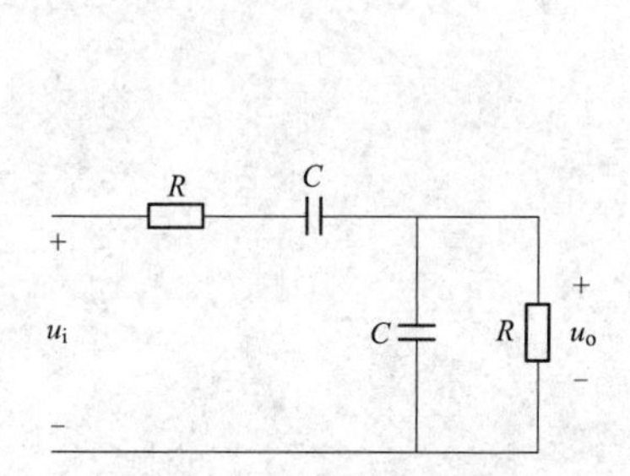

Figure 10 - 1 Series-Parallel *RC* Circuit

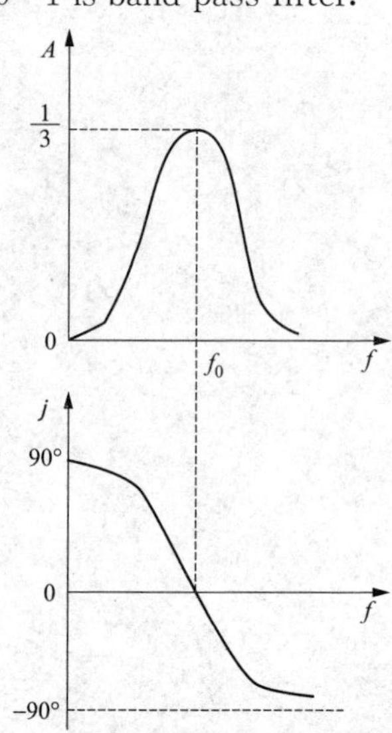

Figure 10 - 2 The Amplitude Response and the Phase Response of Series-Parallel *RC* Circuit

At $\omega = \frac{1}{RC}$, we have $A(\omega) = \frac{1}{3}$, $\varphi(\omega) = 0°$, u_O and u_i are in phase, resonance occurs at the resonant frequency $f_0 = \frac{1}{2\pi RC}$. In other words, when the signal frequency is f_0, the output voltage u_O and the input voltage u_i of series-parallel *RC* circuit are in phase, the value of u_O is 1/3 of u_i. This characteristic is called the frequency selectivity of series-parallel *RC* circuit, and this circuit is also known as Wien bridge.

Measuring the amplitude response: keep the output voltage of the signal generator (the input voltage of *RC* network) unchanged, change frequency f, monitor U_i with the oscilloscope, measure the corresponding output voltage U_O of the *RC* network, and calculate the ratio $A=U_O/U_i$. Take f as the horizontal axis and U_O as the vertical axis, draw the amplitude response curve point by point.

Measuring the phase response: keep the output voltage of the signal generator (the input voltage of *RC* network) unchanged, change frequency f, and observe the waveforms of u_O and u_i with the oscilloscope. The input and output waveforms are shown in Figure 10 - 3. If the delay between the two waveforms is Δt, and the period is T, the phase difference is $\varphi = \frac{\Delta t}{T} \times 360°$. Take f as the horizontal axis and φ as the vertical axis, and draw the phase response curve point by point.

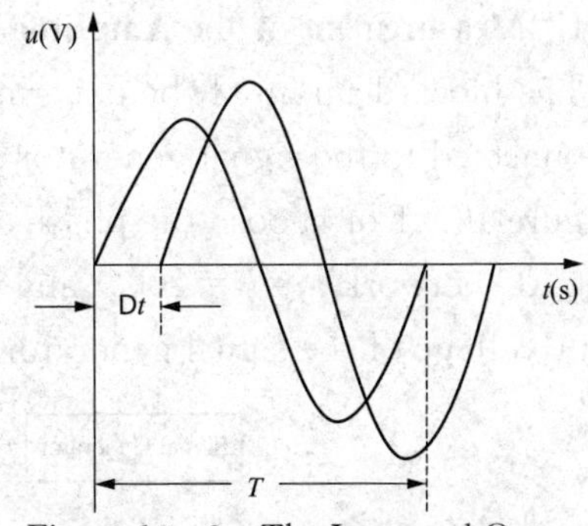

Figure 10 - 3 The Input and Output Waveforms of Series-Parallel *RC* Circuit

The amplitude response of *RC* double—T circuit can be measured using the same method. The *RC* double—T circuit is shown in Figure 10 - 4, and its amplitude response shows band-stop characteristic, as shown in Figure 10 - 5.

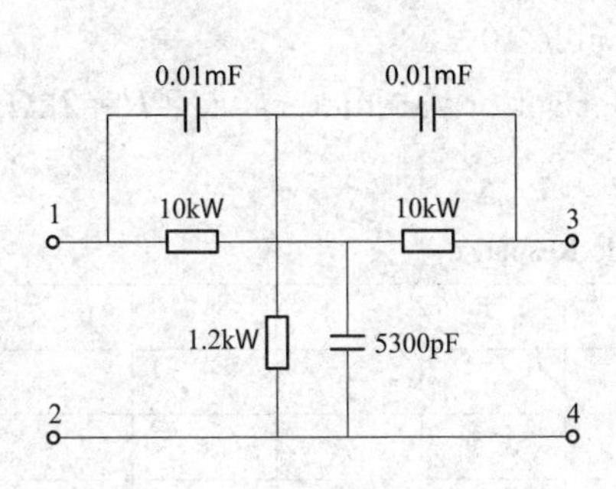

Figure 10 - 4 *RC* Double - T Circuit

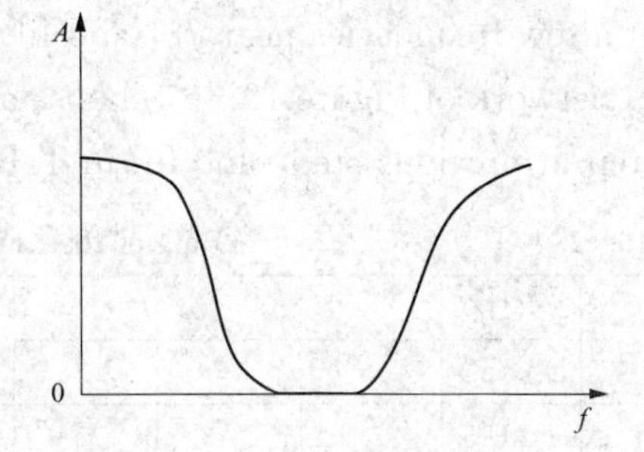

Figure 10 - 5 The Amplitude Response of *RC* Double-T Circuit

- **Equipment**

Equipment is shown in Table 10 - 1.

Table 10 - 1 Equipment

Equipment	Model or Specification	Quantity	Module
Oscilloscope	GDS - 1102A - U	1	
Signal Generator	DG1022U	1	
Experiment Circuit	Phase Selection Circuit	1	NDG - 12
	Double-T Network	1	
Resistor	200Ω	1	NDG - 13
	2kΩ	1	
Capacitor	0. 22μF	1	
	2. 2μF	1	

- **Contents**

1. Measurement of the Amplitude Responses of Series-Parallel *RC* Circuit

The block diagram of the experiment circuit is shown in Figure 10 - 6, where the probes are connected to the signal generator and the oscilloscope. Connect the *RC* network according to Figure 10 - 1 or choose the phase selection circuit from module NDG - 12. The parameters of the *RC* network are $R=200\Omega$ and $C=2.2\mu F$. The input voltage u_i is the sinusoidal wave output voltage of the signal generator, adjust the amplitude of the output signal to $U_i = 2V$.

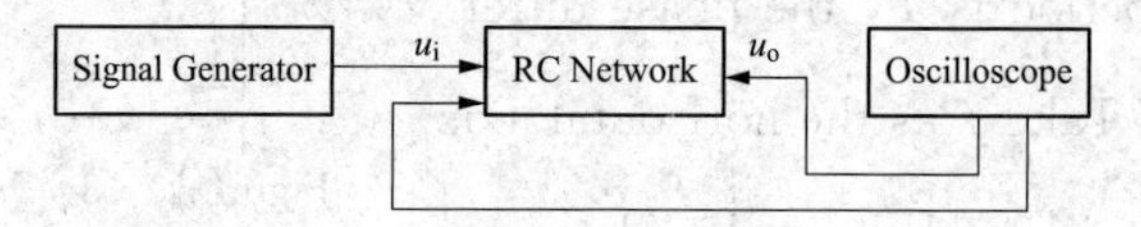

Figure 10 - 6 The Black of Diagram of the Measurement of the Frequency Response of Series-Paraller *RC* Circuit

Change the frequency f of the output sinusoidal wave voltage of the signal generator, and keep $U_i=2V$ constant by monitoring it with the oscilloscope, and measure the output voltage U_O. This can be achieved by measuring the frequency corresponding to $A=1/3$ and then U_O at a few frequencies near f_0, and fill in Table 10 - 2.

In *RC* network of Figure 10 - 6, choose another set of parameters, i. e. $R=2k\Omega$ and $C=0.22\mu F$, repeat previous step, and fill in Table 10 - 2.

Table 10 - 2 Data of the Amplitude Response

$R=2k\Omega$, $C=0.22\mu F$	f (Hz)									
	U_O (V)									
$R=200\Omega$ $C=2.2\mu F$	f (Hz)									
	U_O (V)									

2. Measurement of the Phase Responses of Series-Parallel *RC* Circuit

The experiment circuit is shown in Figure 10 - 6. The principles are stated previously, the experiment procedure is the same as the one in step 1, and fill in Table 10 - 3.

Table 10 - 3 **Data of the Phase Responses**

$R=200\Omega$, $C=2.2\mu F$	f (Hz)									
	T (ms)									
	Δt (ms)									
	φ									
$R=2k\Omega$ $R=1k\Omega$ $C=0.1\mu F$	f (Hz)									
	T (ms)									
	Δt (ms)									
	φ									

3. Measurement of the Amplitude Responses of *RC* Double-T Circuit

The experiment circuit is shown in Figure 10 - 6, and connect the *RC* network according to Figure 10 - 4, or choose the double-T network circuit from module NDG-12. The procedure of the experiment is the same as step 1. Fill the experiment data in a self-made data table.

● **Notes**

In consideration of the influence of the internal resistance of the signal generator on the voltages of the circuit, adjust the value and frequency of the output voltage simultaneously to keep the input voltage constant.

● **Questions**

1. Estimate the resonant frequencies corresponding to the two sets of parameters of the Series-parallel *RC* circuits according to the circuit parameters.

2. Deduce the mathematical expressions of the amplitude response and the phase response of the series-parallel *RC* circuits.

3. What is the frequency selectivity of series-parallel *RC* circuits? What is the relationship between the output and the input of the circuit when the frequency is equal to the resonant frequency?

4. Try to analyze the amplitude response of *RC* double-T circuit qualitatively.

● **Experiment Report**

1. Draw the two sets of the amplitude response and the phase response of series-parallel *RC* circuits using the data of Table 10 - 2 and Table 10 - 3. Find the resonant frequency and the maximum value point of the amplitude response, and compare them with the calculated theoretical value.

2. Design a Wien bridge with a resonant frequency of 1 kHz, and explain its frequency selectivity.

3. Draw the amplitude response of *RC* double-T circuit using the data in experiment 3, and explain the characteristic of this amplitude response.

实验 11　使用交流仪表测定交流电路等效参数

一、实验目的

(1) 学习用交流数字仪表（电压表、电流表、功率表、功率因数表）测量交流电路的电压、电流、功率、功率因数的方法。

(2) 掌握由三表法测得数据，计算出电路等效参数的方法。

(3) 掌握由串联、并联电容判别阻抗性质的方法。

(4) 观测电感、电容元件电压与电流的相位关系。

二、实验原理

(1) 要测量正弦交流电路中各个元件的参数值，可以用交流电压表、交流电流表及功率表分别测量出元件两端的电压 U，流过该元件的电流 I 和它所消耗的功率 P，然后通过计算得到所求各参数值，这种方法称为三表法，是用来测量 50Hz 交流电路参数的基本方法。计算的基本公式如下。

电阻元件的电阻 $R=\frac{U_R}{I}$ 或 $R=\frac{P}{I^2}$

电感元件的感抗 $X_L=\frac{U_L}{I}$，电感 $L=\frac{X_L}{2\pi f}$

电容元件的容抗 $X_C=\frac{U_C}{I}$，电容 $C=\frac{1}{2\pi f X_C}$

串联电路复阻抗的模 $|Z|=\frac{U}{I}$，阻抗角 $\varphi=\arctan\frac{X}{R}$

其中：等效电阻 $R=\frac{P}{I^2}$，等效电抗 $X=\sqrt{|Z|^2-R^2}$。

本次实验电阻元件选用白炽灯（非线性电阻）。电感线圈选用日光灯镇流器，由于镇流器线圈的金属导线具有一定电阻，故镇流器可以表示为一个电感和电阻的串联。电容器一般可认为是理想的电容元件。

(2) 电路功率用功率表测量。功率表是一种电动式仪表，其中电流线圈与负载串联（具有两个电流线圈，可串联或并联，以便得到两个电流量程），而电压线圈与电源并联，电流线圈和电压线圈的同名端（标有 * 号端）必须连在一起，如图 11-1 所示。本实验使用数字式功率表，连接方法与电动式功率表相同，电压、电流量程分别选 450V 和 3A。

在图 11-2 电路中，负载的有功功率 $P=UI\cos\varphi$，其中 $\cos\varphi$ 为功率因数，功率因数角 $\varphi=\arctan\frac{X_L-X_C}{R}$，且 $-90°\leqslant\varphi\leqslant 90°$。

当 $X_L>X_C$，$\varphi>0$，$\cos\varphi>0$，感性负载。

当 $X_L<X_C$，$\varphi<0$，$\cos\varphi>0$，容性负载。

当 $X_L=X_C$，$\varphi=0$，$\cos\varphi=1$，电阻性负载。

可见，功率因数的大小和性质由负载参数的大小和性质决定。

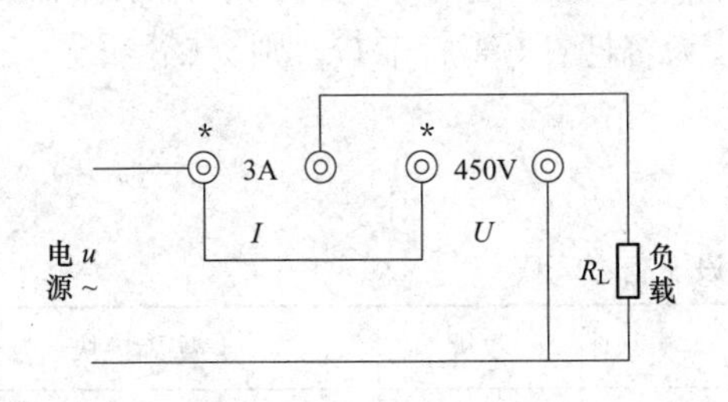

图 11－1　功率表接线图

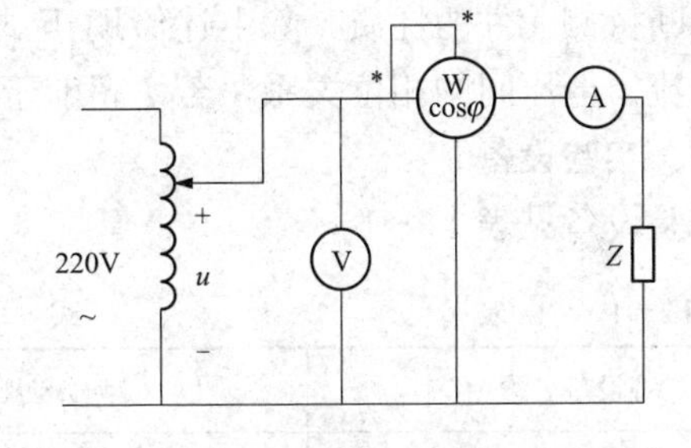

图 11－2　三表法测量交流电路等效参数的基本实验电路

(3) 阻抗性质可由在被测元件两端并联电容或串联电容的方法来加以判别，方法与原理如下：

1）在被测元件两端并联一只适当容量的试验电容，若串接在电路中电流表的读数增大，则被测阻抗为容性，电流减小则为感性。

在图 11－3(a) 中，Z 为待测定的元件，C'为并联电容器。图 11－3(b) 图是图11－3(a) 的等效电路，图中 G、B 为待测阻抗 Z 的电导和电纳，B'为并联电容 C' 的电纳。在端电压有效值不变的条件下，按下面两种情况进行分析：

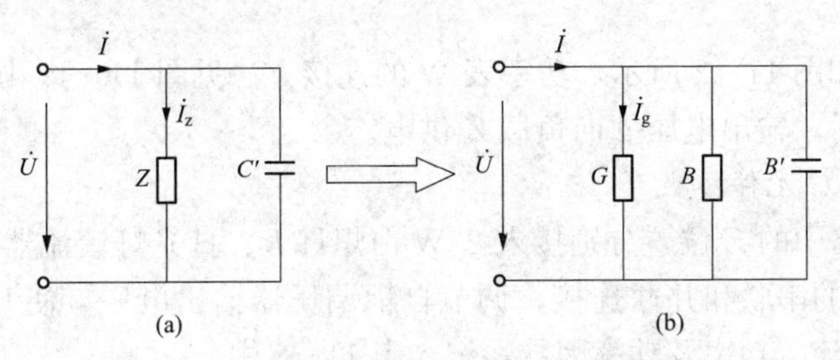

图 11－3　通过并联电容的方式判别阻抗性质

a. 设 $B+B'=B''$，若 B'增大，B''也增大，则电路中电流 I 将单调地上升，故可判断 B 为容性元件。

b. 设 $B+B'=B''$，若 B'增大，而 B''先减小而后再增大，电流 I 也是先减小后上升，如图 11－4 所示，则可判断 B 为感性元件。

由上分析可见，当 B 为容性元件时，对并联电容 C'值无特殊要求；而当 B 为感性元件时，$B'<|2B|$ 才有判定为感性的意义。$B'>|2B|$ 时，电流单调上升，与 B 为容性时相同，并不能说明电路是感性的。因此 $B'<|2B|$ 是判断电路性质的可靠条件，由此得并联电阻的条件为 $C<\left|\frac{2B}{\omega}\right|$。

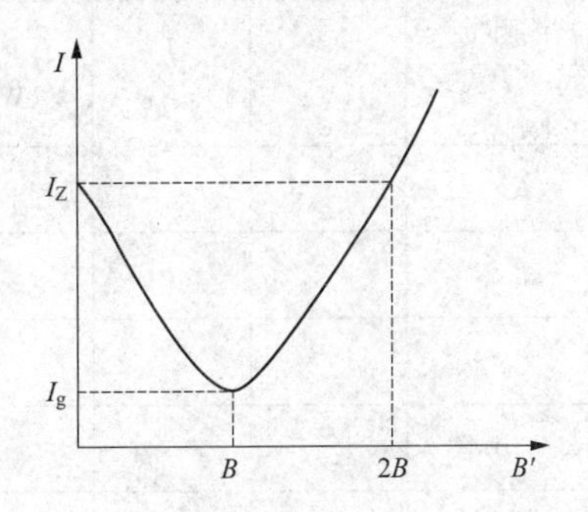

图 11－4　被测元件为感性时电流与并联电纳的变化关系

2）与被测元件串联一个适当容量的试验电容，若被测阻抗的端电压下降，则判为容性，端压上升则为感性，判定条件为 $\frac{1}{\omega C'}<|2X|$，式中 X 为被测阻抗的电抗值，C'为串联试验电容值，此关系式可自行证明。

判断待测元件的性质，除上述借助于试验电容 C' 测定法外，还可以利用示波器观测该元件电流、电压间的相位关系，若 I 超前于 U，为容性；I 滞后于 U，则为感性。

三、实验设备

实验设备见表 11－1。

表 11－1　　实　验　设　备

设备名称	型号与规格	数量	实验模块
双踪示波器	GDS－1102A－U	1	
智能仪表	0～500V 0～3A	3	NDG－01
交流电源	0～450V 三相/0～250V 单相		QS－DYD3
白炽灯泡	25W	3	NDG－10
电感	日光灯镇流器	1	NDG－08
电容	4.3μF/500V	1	

四、实验内容

基本实验电路如图 11－2 所示，功率表 W 的连接方法见图 11－1。图中 220V 交流电源经自耦调压器调压后，输出电压 u 向负载 Z 供电。

1. 测量 R、L、C 元件参数

将图 11－2 电路中的 Z 位置分别接入 25W 白炽灯 R、日光灯镇流器 L 、4.3μF 电容器 C，以及 L、C 元件的串联和并联连接，调节自耦调压器输出电压，使 U 为 150V（用电压表测量），并测量电流、功率及功率因数，记入表 11－2 中。

表 11－2　　三表法测量交流电路等效参数实验数据

被测阻抗	测量值				计算值		电路等效参数		
	U (V)	I (A)	P (W)	$\cos\varphi$	Z (Ω)	$\cos\varphi$	R (Ω)	L (mH)	C (μF)
R									
L									
C									
L、C 串联									
L、C 并联									

2. 验证用串联、并联试验电容法判别负载性质的正确性

实验线路同图 11－2，但不必接功率表，按表 11－3 内容进行测量和记录。

3. 观测元件电流、电压间的相位关系

将图 11－2 中调压器输出电压 u 调至 30V，去掉电流表和功率表，如图 11－5(a) 所示。

分别将 Z 位置接入白炽灯 R 与镇流器 L、白炽灯 R 与 4.3μF 电容 C 的串联，如图 11－5（b）、（c）所示。由于 u_R 与电流 i 同相位，在两电路中分别使用示波器观测 u_R 与 u_L、u_R 与 u_C 的相位关系，即可观察到感性与容性元件上电流与电压的相位关系。

表 11－3　串联、并联试验电容法判别负载性质实验数据

被测元件	串联 1μF 电容		并联 1μF 电容	
	串联前端电压（V）	串联后端电压（V）	并联前电流（A）	并联后电流（A）
R（三只 25W 白炽灯）				
C（4.3μF）				
L（1H）				
L、C 串联				
L、C 并联				

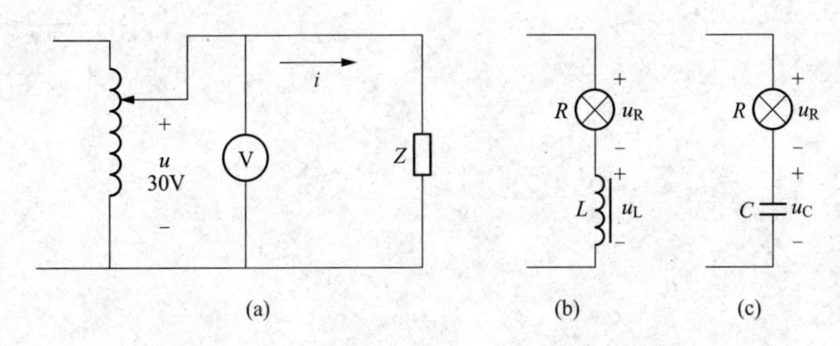

图 11－5　双侧元件电流、电压间的相位关系

（a）基本电路图；（b）观测 u_R 与 u_L；（c）观测 u_R 与 u_C

五、注意事项

（1）自耦调压器在接通电源前，必须将其手柄置在零位上，调节时，使其输出电压从零开始缓慢升高。每次改接实验负载或实验完毕，都必须先将其手柄调回零位，再断电源。必须严格遵守这一安全操作规程。

（2）计算表 11－2 等效参数时，注意根据电路性质（容性或感性）计算对应参数。

（3）通常功率表不单独使用，要有电压表和电流表监测，使电压表和电流表的读数不超过功率表电压和电流的量程。

（4）务必注意示波器输入电压不能过高。

六、思考题

（1）在 50Hz 的交流电路中，测得一只铁芯线圈的 P、I 和 U，如何计算得它的电阻值及电感量？

（2）如何用串联电容的方法来判别阻抗的性质？试用 I 随 X'_C（串联容抗）的变化关系作定性分析，证明串联试验时，C' 满足 $\frac{1}{\omega C'} < |2X|$。

七、实验报告

（1）根据表 11－2 的数据，计算表 11－2 内阻抗、功率因数及电路等效参数。

（2）根据表 11－3 的测量结果，说明如何判别负载性质。

（3）描绘步骤 3 观测到的 u_R 与 u_L、u_R 与 u_C 的波形，说明元件电压与电流的相位关系。

Experiment 11 Measure Equivalent Parameters of AC Circuit with AC Instruments

● Objectives

1. Learn how to measure the voltage, current, power and power factor in AC circuits with AC digital instruments (voltmeter, ammeter, power meter, power factor meter).

2. Learn how to calculate equivalent parameters of circuit according to the data from three-meter method.

3. Learn how to determine impedance properties from series capacitor or shunt capacitor.

4. Observe the phase relations between the voltages and the current of the inductor and the capacitor.

● Principles

To measure the parameters of each element in a sinusoidal AC circuit, measure the voltage U across the element, the current I through the element and the power P the element dissipated. Then calculate the parameters with U, I and P. This method is called three-meter method. Three-meter method is a basic method for measuring the parameters of 50Hz AC circuits. The basic formulas for calculating are listed as follows.

The resistance of the resistor $R=\frac{U_R}{I}$ or $R=\frac{P}{I^2}$.

The reactance of the inductor $X_L=\frac{U_L}{I}$, inductance $L=\frac{X_L}{2\pi f}$.

The capacitive reactance of the capacitor $X_C=\frac{U_C}{I}$, capacitance $C=\frac{1}{2\pi f X_C}$.

The modulus of complex impedance in series circui $|Z|=\frac{U}{I}$, the impedance angle $\varphi=\arctan\frac{X}{R}$.

The equivalent resistance $R=\frac{P}{I^2}$, the equivalent reactance $X=\sqrt{|Z|^2-R^2}$.

In this experiment, the resistor is a set of incandescent lamps (that are nonlinear resistances). The inductance coil is the ballast of the fluorescent lamp. Due to the resistance of the metal wire of the ballast coil, the ballast can be seen as a series of a resistor and an inductor. Generally, the capacitor can be considered an ideal capacitor.

The power of the circuit can be measured by a power meter. The power meter is an electrodynamic instrument in which the current coil is in series with the load and the voltage coil is in parallel with the power supply. The dotted terminals of the current coil and the voltage coil (marked with * signs) must be connected together, as shown in Figure 11－1. In this experiment, the power meters are digital power meters, and their connection method is the same as the electrodynamic power meter. The ranges of voltage and current of the power meters are 450V and 3A.

In the circuit of Figure 11－2, the active power of the load $P=UI\cos\varphi$, in which $\cos\varphi$ is the power factor, and the power factor angle $\varphi=\arctan\dfrac{X_L-X_C}{R}$, and $-90°\leqslant\varphi\leqslant90°$.

When $X_L>X_C$, $\varphi>0$, $\cos\varphi>0$, the load is inductive.

When $X_L<X_C$, $\varphi<0$, $\cos\varphi>0$, the load is capacitive.

When $X_L=X_C$, $\varphi=0$, $\cos\varphi=1$, the load is resistive.

Thus it can be seen, the value and property of the power factor are determined by the values and properties of the parameters of the load.

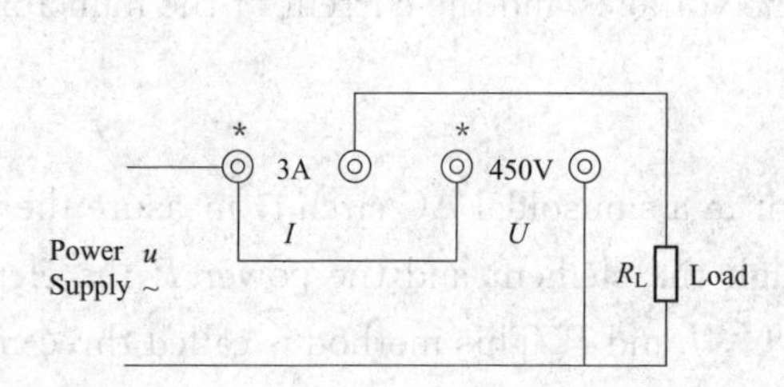

Figure 11－1 Connection of the Power Meter

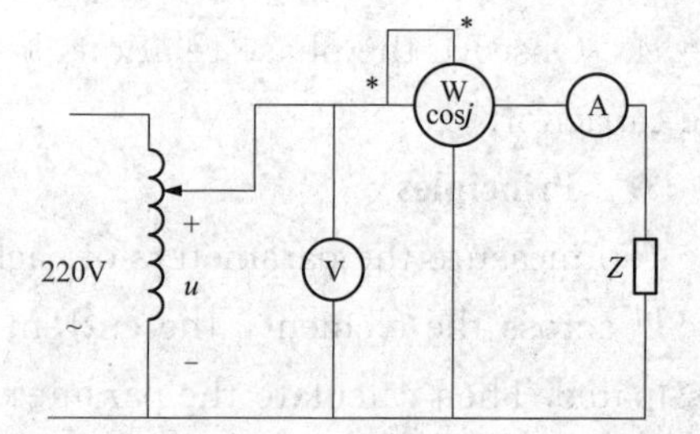

Figure 11－2 The Circuit of Measuring Equivalent Parameters of AC Circuit by Three Meter Method

The property of impedance can be determined by connecting a shunt capacitor or a series capacitor to the measured element, the method and its principle are as follows.

(1) Connect a shunt capacitor of proper capacitance to the measured element. If the reading of the ammeter in series in the circuit increases, the measured impedance is capacitive; if the current decreases, the impedance is inductive.

In Figure 11－3 (a), Z is the element to be measured, C' is the shunt capacitor. The circuit in Figure 11－3 (b) is the equivalent circuit of circuit in Figure 11－3 (a), G and B are the conductance and the susceptance of Z, B' is the susceptance of C'. When the terminal voltage RMS is constant, do the analysis according to the following two cases.

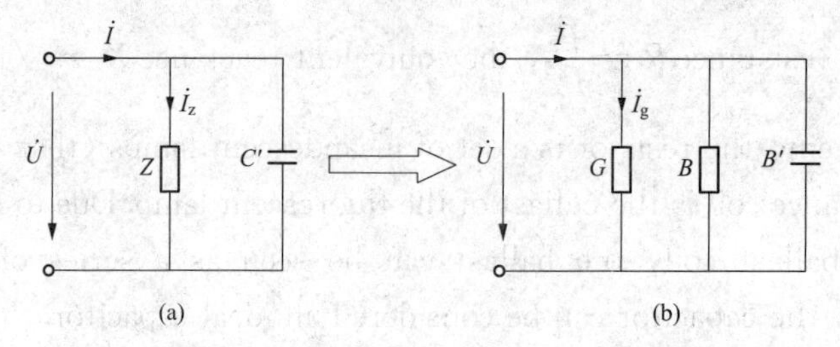

Figure 11－3 Determining the Property of Impedance through a Shunt Capacitor

1) Letting $B+B'=B''$, if B'' increases when B' increases, then the current I of the circuit

increases monotonously, B can be determined as capacitive.

2) Letting $B+B'=B''$, if B'' decreases first then increases, I decreases first then increases too, as shown in Figure 11 - 4, B can be determined as inductive.

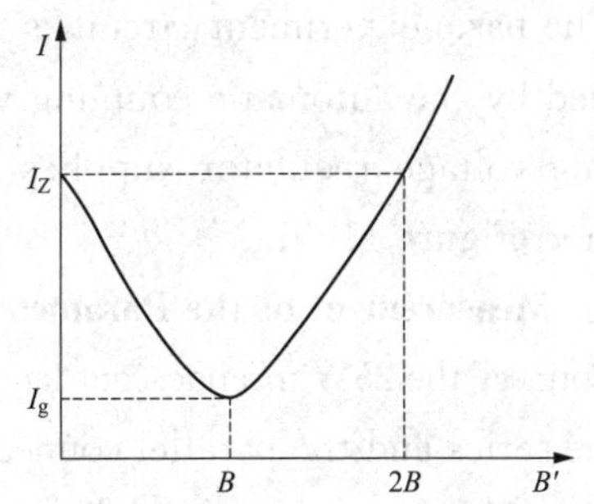

Figure 11 - 4 The Relationship between the Current and the Shunt Susceptance When the Measured Impedance Is Inductive

It can be seen from the above analysis, when B is capacitive, there is no special requirement for the shunt capacitance; but when B is inductive, the determination only makes sense when $B'<|2B|$. When $B'>|2B|$, the current increases monotonously, just the same as B is capacitive, it can't be determined that the circuit is inductive. Therefore $B'<|2B|$ is a reliable condition for determining the property of circuit, the determinant condition of shunt capacitor is $C<\left|\frac{2B}{\omega}\right|$.

(2) Connect a series capacitor of proper capacitance to the measured element, if the voltage across the measured impedance decreases, the impedance is capacitive; and if the voltage increases, the impedance is inductive. The determinant condition is $\frac{1}{\omega C'}<|2X|$, X is the reactance of the measured impedance, and C' is the capacitance of the series capacitor, try to verify this relational expression.

To determine the property of measured element, besides the determination of using experiment capacitors, the oscilloscope also can be used to observe the phase relationship between the voltage and the current of the measured element. If the current is ahead of the voltage, the element is capacitive; if the current lags behind the voltage, the element is inductive.

● Equipment

Equipment is shown in Table 11 - 1.

Table 11 - 1 **Equipment**

Equipment	Model or Specification	Quantity	Module
Oscilloscope	GDS-1102A-U	1	
Smart Meter	0~500V 0~3A	3	NDG-01
AC Power supply	0~450V Three-Phase 0~250V Single-Phase		QS-DYD3
Incandescent Lamp	25W	3	NDG-10
Inductor	Ballast of The Fluorescent Lamp	1	NDG-08
Capacitor	4.3μF/500V	1	NDG-08

- **Contents**

The basic experiment circuit is shown in Figure 11 - 2, the 220V AC power supply is modified by the automatic coupling voltage regulator. The output voltage u of the automatic coupling voltage regulator supplies the load Z. The connection of the power meter W is shown in Figure 11 - 1.

1. Measurement of the Parameters of *R*, *L*, *C*

Connect the 25W incandescent lamp R, the fluorescent light ballast L, the 4.7μF capacitor C, and the series and the parallel connection of L and C to the position of Z in the circuit shown in Figure 11 - 2 respectively. Adjust the output of the automatic coupling voltage regulator to 150V (measured by a voltmeter). Measure the current, the power and the power factor, and fill in Table 11 - 2.

Table 11 - 2 Data of Measuing Equivalent Parameters of AC Circuit by Three-Meter Method

Measured Impedance	Measured Values				Calculated Values		Calculated Equivalent Parameters		
	U (V)	I (A)	P (W)	$\cos\varphi$	Z (Ω)	$\cos\varphi$	R (Ω)	L (mH)	C (μF)
R									
L									
C									
L, C in Series Connection									
L, C in Parallel Connection									

2. Verify the Correctness of Determining the Properties of the Loads with a Shunt Capacitor or a Series Capacitor

The experiment circuit is shown in Figure 11 - 2 but the power meter is not needed. Measure the data in Table 11 - 3 and fill in the table.

3. Observe the Phase Relationship between the Voltage and the Current of the Measured Element

Adjust the output voltage u of the automatic coupling voltage regulator in Figure 11 - 2 to 30V, remove the ammeter and the power meter, as shown in Figure 11 - 5(a). Connect the series of incandescent lamp R and the ballast L, and the series of incandescent lamp R and the 4.3μF capacitor C to the position of Z separately, as shown in Figure 11 - 5(b) and (c). Due to the same phase of u_R and the current i, observe the phase relationships between u_R and u_L, and u_R and u_C with the oscilloscope, the phase relationships between the current and the voltages of the inductive and capacitive elements can be observed.

Table 11 - 3 Data of Determining the Properties of the Loads With a Shunt Capacitor or a Series Capacitor

Measured Element	Connect 1μF Capacitor in Series		Connect 1μF Capacitor in Parallel	
	Voltage Before Connection	Voltage After Connection	Current Before Connection	Current After Connection
R (Three 25W Incandescent Lamps)				
C (4.3μF)				
L (1 H)				
L, C in Series Connection				
L, C in Parallel Connection				

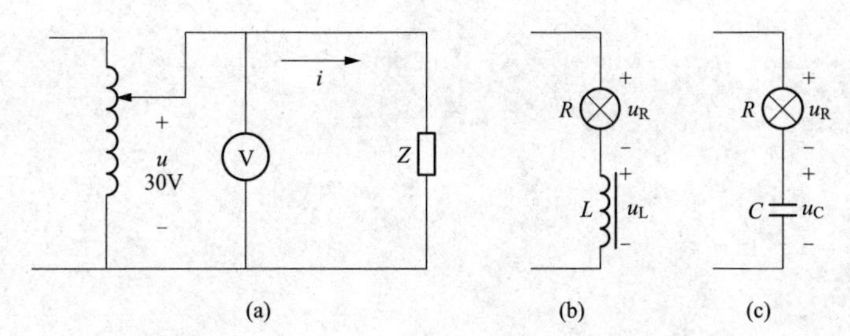

Figure 11 - 5 Observe the Phase Relationship between the Voltage and the Currrent of the Measured Element

(a) The Basic Circuit; (b) u_R and u_L; (c) u_R and u_C

● Notes

1. The handle of the automatic coupling voltage regulator must be in zero position before the power is on, the output voltage shall be increased slowly from zero when the automatic coupling voltage regulator is being adjusted. Every time when the load is changed or the experiment is finished, turn the handle to zero position before the power is off. This safety operation specification must be observed.

2. Calculate the corresponding parameters in Table 11 - 2 according to the properties of the circuit (capacitive or inductive).

3. Usually the power meter isn't used separately, there shall be voltmeter and ammeter for monitoring, the readings of the voltmeter and ammeter shall not be exceed the voltage and current ranges of the power meter.

4. It must be noted that the input voltage of the oscilloscope cannot be too high.

● Questions

1. In a 50Hz AC circuit, P, I and U of an iron core coil have been measured, how to calculate its resistance and inductance?

2. How to determine the property of the load with the method of connecting a series capacitor to the load? Try to analyze the change of current I with the change of X'_C(series capacitive reactance) qualitatively, prove C' satisfies $\frac{1}{\omega C'} < |2X|$ in the series experiment.

- **Experiment Report**

1. Calculate the impedance, power factor and the equivalent parameters in Table 11-2 using the measured data of step 1.

2. Explain how to determine the property of the load using the measuring result in Table 11-3.

3. Draw the waveforms of u_R and u_L, u_R and u_C observed in experiment step 3, explain the phase relationships of voltages and current of elements.

实验 12 正弦稳态交流电路相量的研究

一、实验目的

(1) 研究正弦稳态交流电路中电压、电流相量之间的关系，加深对阻抗、阻抗角及相位差等概念的理解。

(2) 掌握 RC 串联电路的相量轨迹及其作移相器的应用。

(3) 理解改善电路功率因数的意义并掌握其方法。掌握日光灯线路的接线。

二、实验原理

(1) 在单相正弦交流电路中，可用交流仪表测得各支路中的电流值与回路各元件两端的电压值，它们之间的关系满足相量形式的基尔霍夫定律，即$\sum \dot{I}=0$ 和$\sum \dot{U}=0$。

(2) 如图 12－1 所示的 RC 串联电路，在正弦稳态信号 $\dot{U}$ 的激励下，$\dot{U}_R$ 与 $\dot{U}_C$ 保持有 90° 的相位差，即当阻值 R 改变时，$\dot{U}_R$ 的相量轨迹是一个半圆，$\dot{U}$、$\dot{U}_C$ 与 $\dot{U}_R$ 三者形成一个直角形的电压三角形。R 值改变时，可改变 φ 角的大小，从而达到移相的目的。

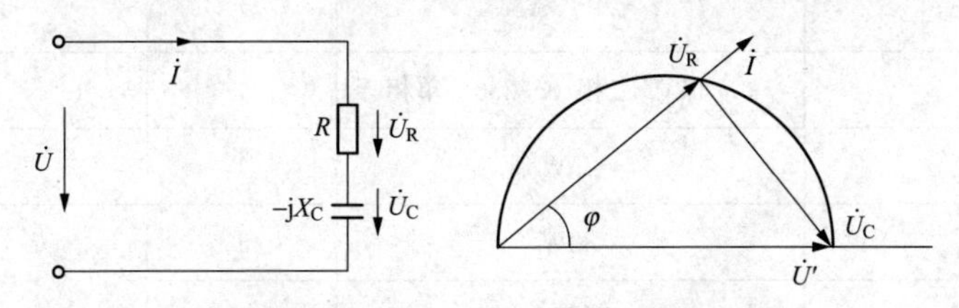

图 12－1 RC 串联电路及其电压相量图

(3) 功率因数 $\cos\varphi=\dfrac{P}{UI}$ 表明，当负载两端电压 U 和消耗的有功功率 P 都不变时，如果功率因数 $\cos\varphi$ 提高，则电流 I 相应减少，视在功率 $S=UI$ 也相应减少，线路损耗 $\Delta P=I^2R_L$ 也随着减少，因此也就提高了传输效率（$\eta=\dfrac{P}{P+\Delta P}$），这就是提高功率因数的实际意义。

(4) 日光灯的构造及其工作原理。日光灯电路由灯管、镇流器和启辉器三部分组成。灯管是一根内壁均匀涂有荧光物质的细长玻璃管，管内充有稀薄的惰性气体和水银蒸汽，在管的两端装有钨灯丝，涂于灯丝上的氧化物受热后易于发射电子的氧化物。镇流器是一个带有铁芯的电感线圈。启辉器由一个辉光管、一个小电容器、两个电极（倒 U 型双层金属动片和固定金属片）组成，如图 12－2 所示。小电容器并联在两个电极之间，其作用是消除火花放电对附近无线设备的影响，并与镇流器组成振荡电路，延迟日光灯灯丝的预热时间，有利于启动。

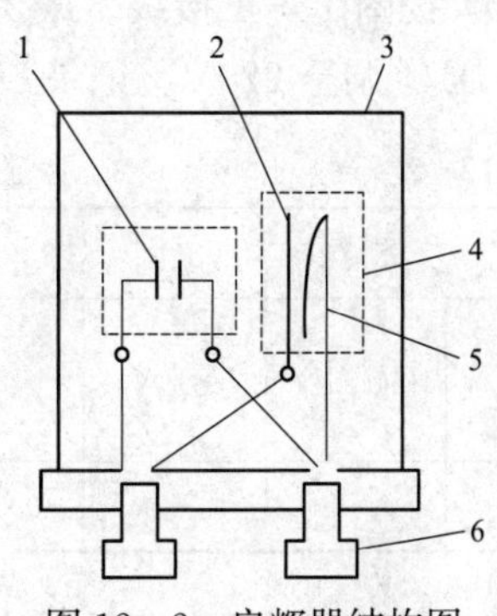

图 12－2 启辉器结构图

1—电容器；2—固定片；3—圆柱形外壳；4—辉光管；5—倒 U 型双层金属动片；6—插脚

当电源接通时，电源电压全部加在启辉器的辉光管两个电极之间，使之放电，放电产生的热量使倒U型双层金属片受热。由于组成双层金属片的两种材料的膨胀系数不同，金属片变形伸展，两电极接通，接通后电极间电压为零，辉光放电停止，倒U型双层金属片因温度下降而复原，两电极脱开。由于回路中电流突然被切断，在镇流器两端产生一个比电源电压高得多的感应电压，此感应电压与电源电压叠加在灯管的两端，使灯丝发热，发射出大量电子，令灯管内惰性气体电离生热，热量令管内水银全部转变为蒸汽，随之水银蒸汽也被电离，发射出强烈的紫外线。在紫外线的激发下，灯管内壁的荧光粉发出接近白色的可见光。

日光灯正常工作时，灯管两端的电压较低，不足以使启辉器再次产生辉光放电。因此启辉器仅在启动过程中起作用，一旦启动完成，便处于断开状态。

三、实验设备

实验设备见表12-1。

表12-1　　实　验　设　备

设备名称	型号与规格	数量	实验模块
智能仪表	0～500V 0～3A	3	NDG-01
交流电源	0～450V三相/0～250V单相		QS-DYD3
白炽灯泡	25W	1	NDG-10
日光灯	30W	1	NDG-08
电容	0.47μF/500V	1	
	1μF/500V	1	
	2.2μF/500V	2	
	4.3μF500V	1	

四、实验内容

(1) 按图12-1接线。R为25W白炽灯泡，电容器C为4.3μF/500V。将自耦调压器输出电压U调至220V。记录U、U_R、U_C值，记入表12-2中，验证电压三角形关系。

表12-2　　验证*RC*串联电路电压三角形关系实验数据

测量值			计算值		
U (V)	U_R (V)	U_C (V)	U'（与U_R，U_C组成直角三角形） ($U'=\sqrt{U_R^2+U_C^2}$)	$\Delta U=U'-U$ (V)	$\Delta U/U$ (%)

(2) 测量日光灯电路。日光灯电路如图12-3所示，L为镇流器，A为日光灯，S为启辉器。u为自耦调压器输出电压，缓慢调节自耦调压器，使其输出电压从零开始增大，直至日光灯启动且刚刚能够持续发光为止，测量功率P，电流I，电压U、U_L、U_A等值，记入表

12－3 启辉值对应栏目内。然后将电压调至 220V，测量上述参数的正常工作值，验证电压、电流相量关系。

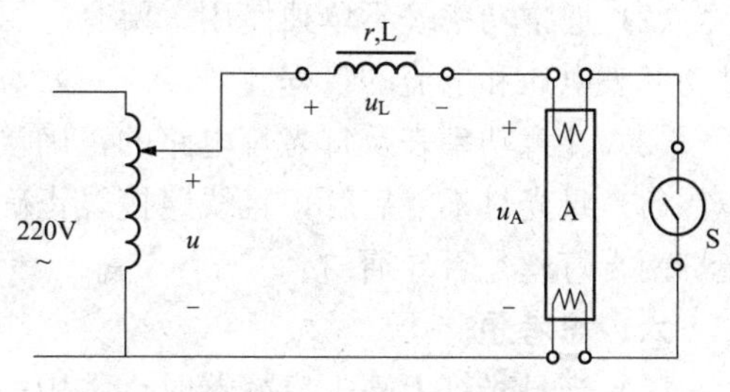

图 12－3　日关灯电路

(3) 电路功率因数的提高。按图 12－4 组成实验电路，C_1～C_N 是补偿电容器，用以改善电路的功率因数（cosφ 值）。调节自耦调压器的输出调至 220V，记录功率表，电压表读数，通过一只电流表和三个电流取样插座分别测得三条支路的电流。根据实验面板所提供的电容，将单个电容或若干个电容的组合并入电路，找到令 I 最小的电容值，将数据填入表 12－4 中，并各取 1～2 组小于或大于此最佳补偿值的电容值，测量数据，填入表中（表格行数如不足请自行扩展）。

表 12－3　　日关灯电路实验数据

测量名称	测量值					计算值	
	P（W）	I（A）	U（V）	U_L（V）	U_A（V）	cosφ	r（Ω）
启辉值							
正常工作值							

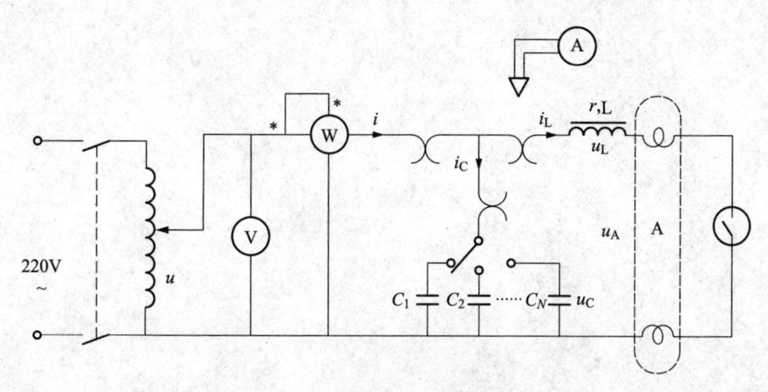

图 12－4　日关灯功率因数补偿电路

表 12－4　　功率因数补偿实验数据

C（μF）	测量值								计算值	
	P（W）	U（V）	U_c（V）	U_L（V）	U_A（V）	I（A）	I_C（A）	I_L（A）	I'（A）	cosφ

五、注意事项

(1) 自耦调压器在接通电源前，必须将其手柄置在零位上，调节时，使其输出电压从零开始缓慢升高。每次改接实验负载或实验完毕，都必须先将其手柄调回零位，再断电源。在并联电容实验内容中，应特别注意遵守这一安全操作规程，否则可能引起实验台告警。

（2）通常功率表不单独使用，要有电压表和电流表监测，使电压表和电流表的读数不超过功率表电压和电流的量程。

（3）注意功率表及日光灯电路的正确接线，加电前须经指导教师检查。

（4）日光灯不能启辉，且线路接线正确时，应检查：①启辉器及其接触是否良好；②日光灯两端灯丝是否导通。

六、思考题

（1）当日光灯上缺少启辉器时，常用一根导线将启辉器插座的两端短接一下，然后迅速断开，使日光灯点亮；或用一只启辉器去点亮多只同类型的日光灯，这样做的原理是什么？

（2）为了提高电路的功率因数，常在感性负载上并联电容器，此时增加了一条电流支路，试问电路的总电流是增大还是减小，此时感性元件上的电流和功率是否改变？

（3）为什么只采用并联电容器法提高线路功率因数，而不用串联法？所并联的电容器是否越大越好？

七、实验报告

（1）完成表 12-2 内计算并画出电压相量图。

（2）完成表 12-3 内计算，画出各个电压和电流的相量图，利用相量图求出 r，说明各个电压之间的关系。

（3）完成表 12-4 内计算，画出各个电压和电流的相量图，分析说明并联电容对电路的补偿作用。

Experiment 12 Study of the Phasors in a Sinusoidal Steady-State AC Circuit

● **Objectives**

1. Study the relationship between the voltage phasor and current phasor in sinusoidal steady-state AC circuits, acquire a better understanding of impedance, impedance angle and phase difference.

2. Understand the phasor trajectory of *RC* series circuit and the usage of it as phase shifter.

3. Understand the significance the power factor in electric circuit and learn how to improve it. Learn how to connect the fluorescent lamp circuit.

● **Principles**

1. In a single phase sinusoidal AC circuit, measure the current in the branches and the voltages across the elements in the loops, the relationships among them are satisfied Kirchhoff's laws in phasor form, $\sum \dot{I}=0$ and $\sum \dot{U}=0$.

2. In a series *RC* circuit shown in Figure 12－1, $\dot{U}_R$ and $\dot{U}_C$ keep a phase difference of 90° under the excitation of sinusoidal steady signal $\dot{U}$. In other words, when the resistance *R* changes, the phasor trajectory of $\dot{U}_R$ is a semi-circle, $\dot{U}$, $\dot{U}_C$ and $\dot{U}_R$ form a right angled triangle. When *R* changes, the angle φ changes, thus the purpose of phase shift is achieved.

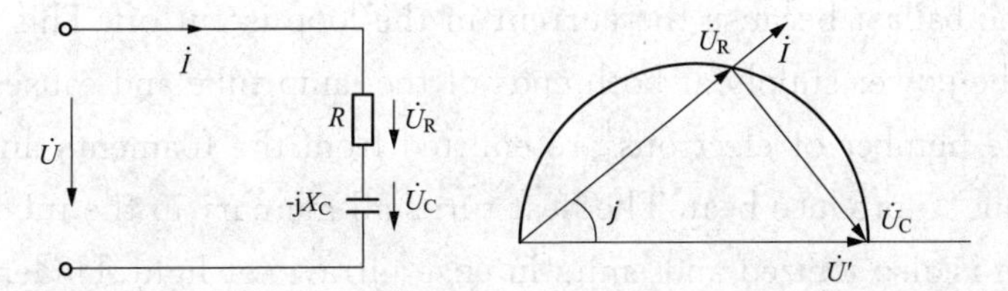

Figure 12－1 A Series *RC* Circuit and Its Phasor Diagram

3. The power factor $\cos\varphi=\dfrac{P}{UI}$, the formula shows when the voltage *U* across the load and the power *P* dissipated by the load are both unchanged. If the power factor $\cos\varphi$ increases, the current *I* decreases correspondingly, the apparent power $S=UI$ decreases correspondingly too, and the line loss $\Delta P=I^2R_L$ decreases. Therefore the transmission efficiency ($\eta=\dfrac{P}{P+\Delta P}$) is improved. This shows the practical significance of improving the power factor.

4. The Structure and Working Principle of the Fluorescent Lamp

The fluorescent lamp circuit consists of three parts, the lamp tube, the ballast and the starter. The fluorescent lamp tube is a slender glass tube with fluorescent substance evenly coated on the inner wall. The tube is filled with thin inert gas and mercury vapor, tungsten filaments are installed at both ends of the tube. Some kind of oxide that is easy to emit electrons after being heated is coated on the filaments. The ballast is an inductor with an iron core. The starter consist of a glow tube, a small capacitor, and two electrodes (an inverted-U-shape moveable dual-layer metal sheet and a fixed metal sheet), as shown in Figure 12 - 2. The small capacitor is in parallel with the two electrodes, the usage of the capacitor is to eliminate the impact of spark discharge on nearby wireless devices, and to make up an oscillating circuit with the ballast, so the preheating time of the fluorescent lamp filaments could be delayed, which beneficial for starting up.

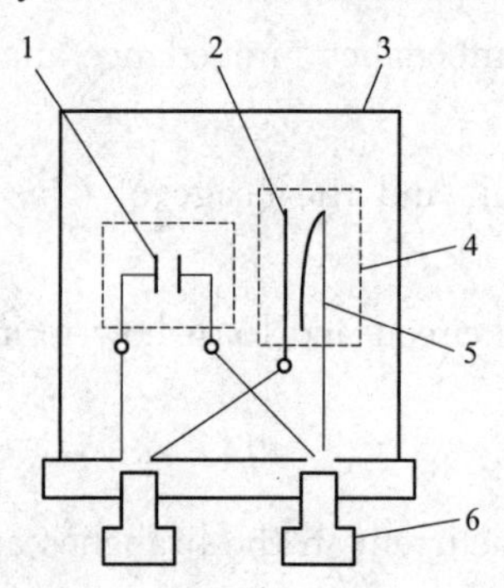

Figure 12 - 2 The Structure of the Starter
1—Capacitor; 2—Fixed Metal Sheet; 3—Cylindrical Shell; 4—Glow Tube; 5—Inverted- U -Shape Moveable Dual-Layer Metal Sheet; 6—Pin

When the power is turned on, the voltage of the power supply is added to the two electrodes of the glow tube of the starter and makes the electrodes discharge. The heat generated by the discharge causes the inverted-U-shape moveable dual-layer metal sheet to be heated. The coefficients of expansion of the two kinds of materials that make up the dual-layer metal sheet are different, and the metal sheet deforms and stretches. The two electrodes are connected. The glow discharge stops when the voltage between the connected electrodes is zero. The inverted-U-shape metal sheet restitutes because of the lowering of temperature, and the two electrodes separate. An induction voltage that is much higher than the power supply voltage is produced at both ends of the ballast because the current in the loop is cut off. This induction voltage superposes with the power supply at both ends of the lamp tube and causes the filaments to be on heat. A large number of electrons are emitted from the filaments and ionize the inert gas in the lamp tube to produce heat. The heat turns all mercury in the tube into steam. Soon the mercury steam is also ionized and emits intense ultraviolet light. Under the excitation of ultraviolet light, the fluorescent powder on the inner wall of the lamp emits near white visible light.

When a fluorescent lamp works normally, the voltage at both ends of the lamp tube is too low to cause glow discharges in the starter again. Therefore the starter operates only in the process of starting, once the starting is completed, it is in a disconnected state.

- **Equipment**

Equipment is shown in Table 12 - 1.

- **Contents**

1. Connect the circuit according to Figure 12 - 1, R is a 25W incandescent bulb, and the capacitor C is 4.7μF/450V. Adjust the output voltage of the automatic coupling voltage regulator to 220V. Measure the values of U, U_R, U_C, and fill in Table 12 - 2. Verify the triangle relation of the voltages.

Table 12 - 1 Equipment

Equipment	Model or Specification	Quantity	Module
Smart Meter	0~500V 0~3A	3	NDG - 01
AC Power Supply	0~450V Three-Phase 0~250V Single-Phase		QS - DYD3
Incandescent Bulb	25W	1	NDG - 10
Fluorescent Lamp	30W	1	NDG - 08
Capacitor	0. 47μF/ 500V	1	
	1μF/500V	1	
	2. 2μF/500V	2	
	4. 3μF/500V	1	

Table 12 - 2 Data of Verifying the Triangle Relation of the Voltages in Series *RC* Circuit

Measured Value			Calculated Value		
U (V)	U_R (V)	U_C (V)	U' (form a right triangle with U_R and U_C) ($U'=\sqrt{U_R^2+U_C^2}$)	$\Delta U=U'-U$ (V)	$\Delta U/U$ (%)

2. The Measurements of the Fluorescent Lamp Circuit

The fluorescent lamp circuit is shown in Figure 12 - 3, L is the ballast, A is the lamp itself, and S is the starter. u is the output voltage of the automatic coupling voltage regulator, slowly increase u from zero until the fluorescent lamp starts and has persistent luminescence. Measure the power P, the current I, and the voltage U, U_L and U_A, and fill them in the "starting value" column of Table 12 - 3. Then adjust the output voltage to 220V, measure the normal working values of the above parameters, verify the relations of phasors of current and voltages.

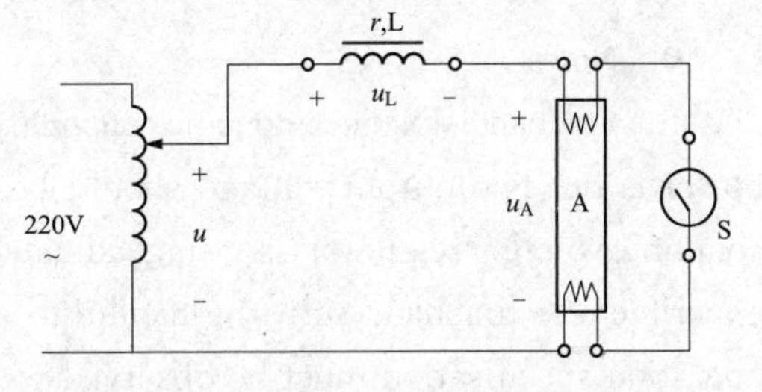

Figure 12 - 3 The Fluore Scent Lamp Circuit

3. The Improvement of the Power Factor

Connect the experiment circuit according to Figure 12 - 4. $C_1 \sim C_N$ are the compensating capacitors for improving the power factor ($\cos\varphi$). Adjust the output voltage of the automatic coupling voltage regulator to 220V, and write down the readings of the power meter and the voltmeter. Measure the current in three branches with an ammeter and three current sampling sockets. Connect the provided capacitors into the circuit respectively or comprehensively until the capacitance that makes I minimum is found, and fill in Table 12 - 4. Then take some sets of capacitances that are smaller or larger than the optimal capacitance, measure the data and fill in the table. If the rows in the table are insufficient, extend them.

Table 12 - 3　　Data of the Measurements of the Fluorescent Lamp Circuit

Measured Name	Measured Value					Calculated Value	
	P (W)	I (A)	U (V)	U_L (V)	U_A (V)	$\cos\varphi$	r (Ω)
Starting Value							
Normal Working Value							

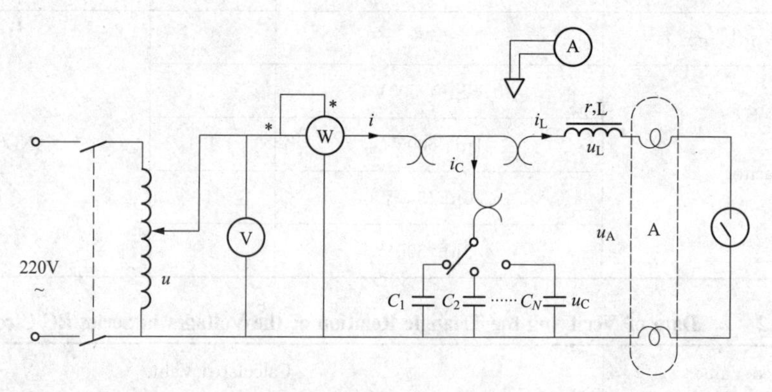

Figure 12 - 4　The Power Factor Compensation Circuit for the Fluorescent Lamp

Table 12 - 4　　Data of the Power Factor Compensation

C (μF)	Measured Value								Calculated Value	
	P (W)	U (V)	U_c (V)	U_L (V)	U_A (V)	I (A)	I_C (A)	I_L (A)	I' (A)	$\cos\varphi$

- **Notes**

1. The handle of the automatic coupling voltage regulator must be in zero position before the power is on. The output voltage should be increased slowly from zero when the automatic coupling voltage regulator is being adjusted. Every time when the load is changed or the experiment is finished, turn the handle to zero position before the power is off. This safety operation specification must be observed.

2. Usually a power meter isn't used separately. There shall be voltmeter and ammeter for monitoring. The readings of the voltmeter and the ammeter shall not exceed the voltage and the current ranges of the power meter.

3. Pay attention to the connections of the power meter and the fluorescent lamp circuit. Let the teacher check the circuit before the power is on.

4. When a fluorescent lamp cannot start and the circuit connection is right, consider the following. ①Is the starter in good condition and good contact? ②Are the filaments at both

ends of the fluorescent lamp in conduction?

- **Questions**

1. When the starter is missing on the fluorescent lamp, people usually connect the both ends of the starter socket with a wire and disconnect it quickly, then the fluorescent lamp starts. Or start multiple fluorescent lamp of the same type with one starter. What are the principles of these acts?

2. To improve the power factor of the circuit, a shunt capacitor is used on an inductive load. At this time, a circuit branch is added to the circuit. Does the total current in the circuit increase or decrease? Do the current and power on the inductive elements change?

3. Why do we use the shunt capacitors to improve the power factor of circuits rather than series capacitors? Are the bigger the shunt capacitors the better?

- **Experiment Report**

1. Finish the calculation of Table 12 - 2 and draw phasor diagrams of the voltages.

2. Finish the calculation of Table 12 - 3 and draw the phasor diagrams of the voltages and current. Explain the relationships among the voltages.

3. Finish the calculation of Table 12 - 4 and draw the phasor diagrams of the voltages and current. Analyze and explain the compensation of the capacitor to the circuit.

实验 13　最大功率传输条件的研究

一、实验目的

（1）理解阻抗匹配，掌握最大功率传输的条件。

（2）掌握根据电源外特性设计实际电源模型的方法。

二、实验原理

电源向负载供电的电路如图 13-1 所示，图中 R_S 为电源内阻，R_L 为负载电阻。当电路电流为 I 时，负载 R_L 得到的功率为

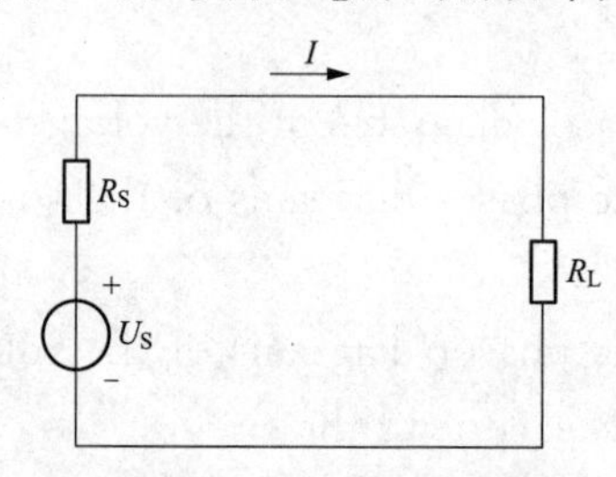

图 13-1　电源向负载供电的电路

$$P_L = I^2 R_L = \left(\frac{U_S}{R_S + R_L}\right)^2 R_L$$

可见，当电源 U_S 和 R_S 确定后，负载得到的功率大小只与负载电阻 R_L 有关。

令 $\frac{dP_L}{dR_L} = 0$，解得当 $R_L = R_S$ 时，负载得到最大功率为

$$P_{Lmax} = \frac{U_S^2}{4R_S}$$

$R_L = R_S$ 称为阻抗匹配，即电源的内阻抗（或内电阻）与负载阻抗（或负载电阻）相等时，负载可以得到最大功率。也就是说，最大功率传输的条件是供电电路必须满足阻抗匹配。

负载得到最大功率时电路的效率 $\eta = \frac{P_L}{U_S I} = 50\%$。

实验中，负载得到的功率用电压表、电流表测量。

三、实验设备

实验设备见表 13-1。

表 13-1　　实　验　设　备

设备名称	型号与规格	数量	实验模块
恒压源	0～30V	1	NDG-02
恒流源	0～500mA	1	
直流电压表	0～200V	1	NDG-03
直流电流表	0～2000mA	1	
电阻	自选		NDG-13
电位器	1kΩ（可选）		

四、实验内容

（1）根据电源外特性曲线设计一个实际电压源模型。已知电源外特性曲线如图 13-2 所示，根据图中给出的开路电压和短路电流数值，计算出实际电压源模型中的电压源 U_S 和内

阻 R_S。实验中，电压源 U_S 选用恒压源的可调稳压输出端，内阻 R_S 选用固定电阻。

（2）测量电路传输功率。用上述设计的实际电压源与负载电阻 R_L 相连，电路如图 13-3 所示，图中 R_L 选用 1kΩ 电位器（或选用实验模块上的固定值电阻），从 0～600Ω 改变负载电阻 R_L 的数值，测量对应的电压、电流，R_L 的数值可用万用表测量，将数据记入表 13-2 中。

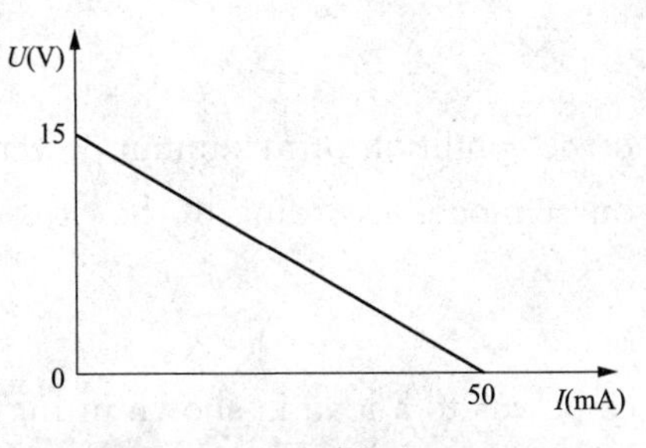

图 13-2　电源外特性曲线

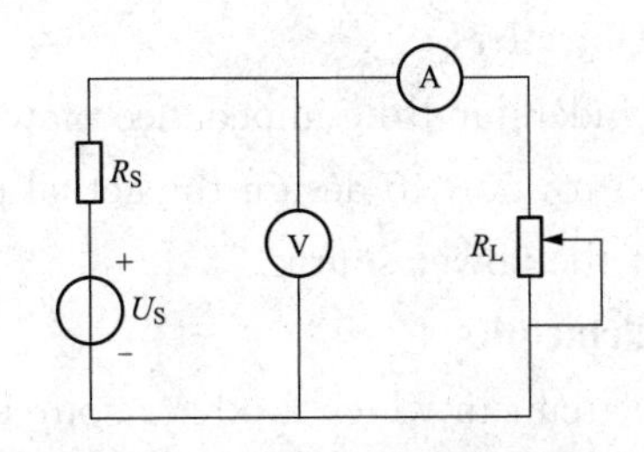

图 13-3　测量电路传输功率

表 13-2　电路传输功率数据

R_L（Ω）							
U（V）							
I（mA）							
P_L（mW）							
η（%）							

五、注意事项

恒压源输出电压根据计算的电压源 U_S 数值进行调整。防止电源短路。

六、思考题

若电压表、电流表前后位置对换，对电压表、电流表的读数有无影响？为什么？

七、实验报告

（1）根据表 13-2 的实验数据，计算出对应的负载功率 P_L，并画出负载功率 P_L 随负载电阻 R_L 变化的曲线，找出传输最大功率的条件。

（2）根据表 13-2 的实验数据，计算出对应的效率 η，指明：①传输最大功率时的效率；②什么时候出现最大效率？由此说明电路在什么情况下，传输最大功率才比较经济、合理。

Experiment 13 Study of Maximum Power Transfer Condition

● Objectives

1. Understand the impedance matching and the condition of maximum power transfer.

2. Learn how to design the actual power source model according to the external characteristic of the power source.

● Principles

The circuit in which a power source supplies power to a load is shown in Figure 13-1, R_S is the internal resistance of the power source, and R_L is the load resistance. I is the current through the circuit, and the power of the load R_L is

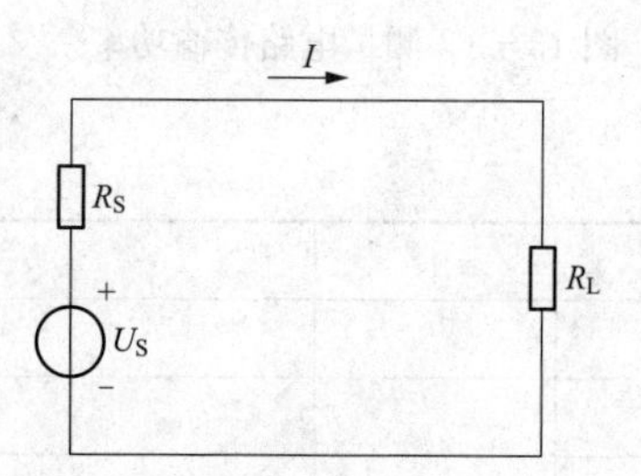

Figure 13-1 The Circuit in Which a Power Source Supplies Power to a Load

$$P_L = I^2 R_L = \left(\frac{U_S}{R_S + R_L}\right)^2 R_L$$

Thus it can be seen, when the power source U_S and its internal resistance R_S are certain, the power of the load is only related to the load resistance R_L.

Letting $\frac{dP_L}{dR_L} = 0$, solution is when $R_L = R_S$, the load gets the maximum power $P_{Lmax} = \frac{U_S^2}{4R_S}$.

$R_L = R_S$ is called impedance matching, which means when the internal impedance (or internal resistance) is equal to the load impedance (or load resistance) and the load gets the maximum power. In other words, the condition for maximum power transfer is the power supply circuit must meet the impedance matching.

The efficiency of the circuit when the load gets the maximum power $\eta = \frac{P_L}{U_S I} = 50\%$.

In the process of the experiment, the power of the load is measured with the voltmeter and the ammeter.

● Equipment

Equipment is shown in Table 13-1.

Table 13-1 **Equipment**

Equipment	Model or Specification	Quantity	Module
Constant Voltage Source	0~30V	1	NDG-02
Constant Current Source	0~500mA	1	

续表

Equipment	Model or Specification	Quantity	Module
DC Voltmeter	0～200V	1	NDG－03
DC Ammeter	0～2000mA	1	
Resistor	Optional		NDG－13
Potentiometer	1kΩ (Optional)		

● **Contents**

1. Design an Actual Voltage Source Model According to the External Characteristic Curve of the Power Source

The external characteristic curve of the power source is shown in Figure 13－2. Calculate the voltage source U_S and the internal resistance R_S of the actual voltage source model using the value of the open circuit voltage and the short circuit current given in the figure. In the experiment, the voltage source U_S is the output of the constant voltage source and the internal R_S is selected from the fixed resistors on the experiment module.

2. Measure the Transmission Power of the Circuit

Connect the designed actual voltage source to the load resistor R_L, as shown in Figure 13－3. R_L is a potentiometer of 1 kΩ, or the fixed resistors on the experiment module. Change the resistance of the load resistor from 0 to 600Ω respectively, and measure the corresponding voltages and current. The resistance of R_L can be measured with the multimeter, fill in Table 13－2.

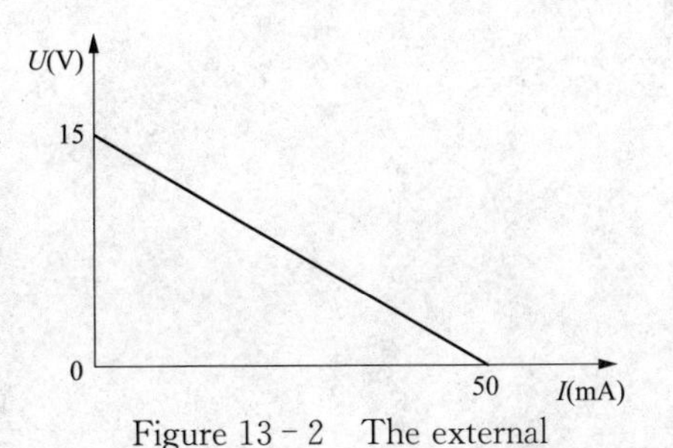

Figure 13－2 The external Characteristic Curve of the Power Source

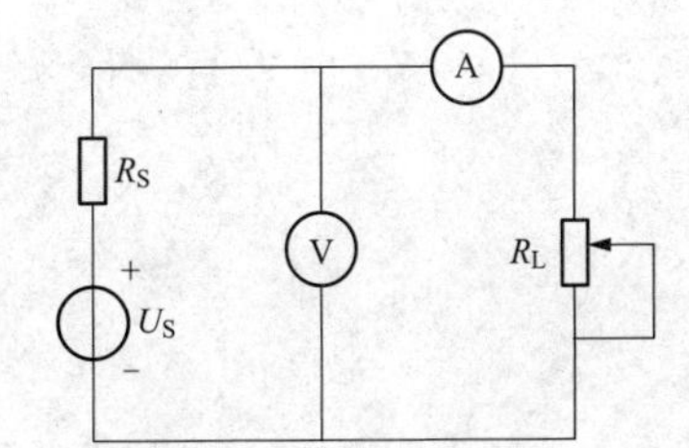

Figure 13－3 Measure the Transmission Power of the Circuit

Table 13－2 Data of Transmission Power of the Circuit

R_L (Ω)							
U (V)							
I (mA)							
P_L (mW)							
η (%)							

● **Notes**

The output of the constant voltage source is adjusted according to the calculated voltage of the voltage source U_S. Beware of the short circuit of the power source.

- **Questions**

Is there any difference on the readings of the voltmeter and ammeter when the front and back positions of them are changed?

- **Experiment Report**

1. Calculate the corresponding load power P_L using the data in Table 13 - 2. Draw the curve of the load power P_L changing with the load resistance R_L and find the maximum power transfer condition.

2. Calculate the corresponding efficiency η using the data in Table 13 - 2, and point out the following. ① The efficiency of transmitting the maximum power. ② When does the maximum efficiency occur? And explain under what conditions the maximum power transmitting is economic and rational.

实验14　互感电路的研究

一、实验目的

（1）学会测定互感线圈同名端、互感系数以及耦合系数的方法。

（2）理解两个线圈相对位置的改变，以及线圈用不同导磁材料时对互感系数的影响。

二、实验原理

一个线圈因另一个线圈中的电流变化而产生感应电动势的现象称为互感现象，这两个线圈称为互感线圈，用互感系数（简称互感）M来衡量互感线圈的这种性能。互感的大小除了与两线圈的几何尺寸、形状、匝数及导磁材料的导磁性能有关外，还与两线圈的相对位置有关。

1. 判断互感线圈同名端的方法

（1）直流法。如图14－1所示，当开关S闭合瞬间，若毫安表的指针正偏，则可断定1、3为同名端；指针反偏，则1、4为同名端。

（2）交流法。如图14－2所示，将两个线圈N_1和N_2的任意两端（如2、4端）连在一起，在其中的一个线圈（如N_1）两端加一个低电压，另一线圈开路，用交流电压表分别测出两互感线圈的端电压U_{12}和U_{34}及两线圈未连接的一组端子之间的电压U_{13}，若U_{13}为两线圈端电压之差，则1、3是同名端；若U_{13}为两线圈端电压之和，则1、4是同名端。

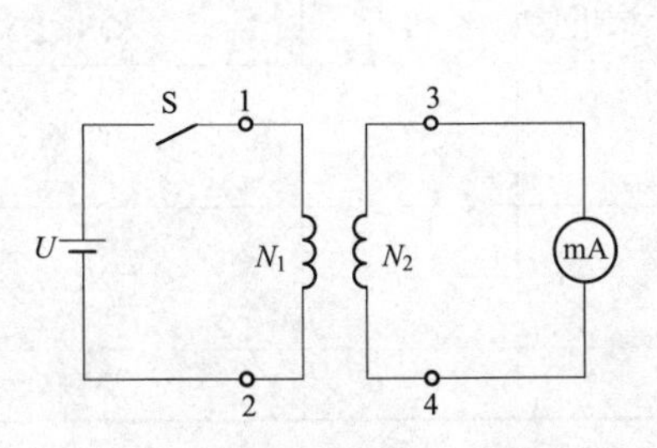

图14－1　直流法判别同名端原理

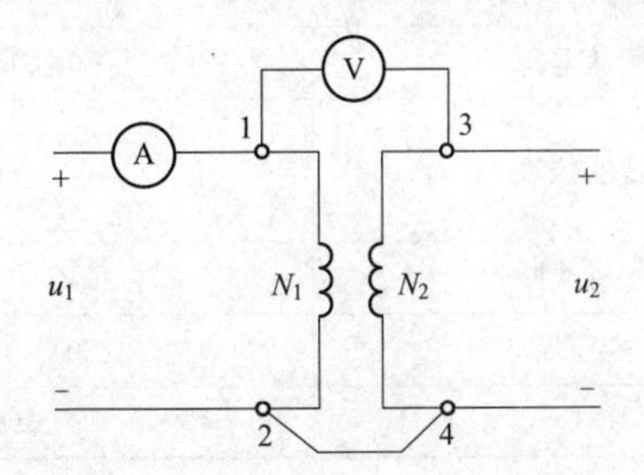

图14－2　交流法判别同名端原理

2. 两线圈互感系数M的测定

在图14－2电路中，去掉两线圈间连线，在互感线圈的N_1侧施加低压交流电压U_1，测出N_1侧电流I_1及N_2侧电压U_2，因此时N_2开路，不存在I_2，故有$U_2=\omega MI_1$，可算得互感系数为

$$M_{21}=\frac{U_2}{\omega I_1}$$

在N_2侧加低压交流电压U_2，N_1开路，测出N_2侧电流I_2及N_1侧电压U_1，有$U_1=\omega MI_2$，可算得互感系数为

$$M_{12}=\frac{U_1}{\omega I_2}$$

应有$M_{21}=M_{12}$，即互感系数M。

3. 耦合系数 k 的测定

两个互感线圈耦合松紧的程度可用耦合系数 k 来表示

$$k = M/\sqrt{L_1 L_2}$$

式中：L_1为 N_1线圈的自感系数；L_2为 N_2线圈的自感系数。

它们的测定方法为先在 N_1侧加低压交流电压 U_1，测出 N_2侧开路时 N_1的电流 I_1；然后再在 N_2侧加电压 U_2，测出 N_1侧开路时 N_2的电流 I_2，有

N_1侧加电压时，有 $\dot{U}_1 = (R_1 + j\omega L_1)\dot{I}_1 \pm j\omega M\dot{I}_2$。

N_2侧加电压时，有 $\dot{U}_2 = (R_2 + j\omega L_2)\dot{I}_2 \pm j\omega M\dot{I}_1$。

因 N_1侧加电压时 N_2侧开路，N_2侧加电压时 N_1侧开路，可知：

N_1侧加电压时，有 $\dot{U}_1 = (R_1 + j\omega L_1)\dot{I}_1$。

N_2侧加电压时，有 $\dot{U}_2 = (R_2 + j\omega L_2)\dot{I}_2$。

R_1、R_2分别为电感线圈 N_1、N_2电阻，在 N_1、N_2未接入电路的状态下，用万用表直接测量出 R_1、R_2，即可分别求出自感 L_1和 L_2。当已知互感系数 M，便可算得 k 值。

三、实验设备

实验设备见表 14-1。

表 14-1 实 验 设 备

设备名称	型号与规格	数量	实验模块
恒压源	0～30V	1	NDG-02
直流电压表	0～200V	1	NDG-03
直流电流表	0～2000mA	1	
交流电源	0～450V 三相/0～250V 单相	1	QS-DYD3
智能仪表	0～500V 0～3A	3	NDG-01
互感线圈	30V/0.5A	1 对	NDG-09
变压器	220V/36V	1	
电阻	200Ω	1	NDG-13

四、实验内容

1. 测定互感线圈的同名端

（1）直流法。按图 14-3 连接电路，电路中交直流设备通过电流取样插座进行连接。将 N_1端子 1、2 接入直流恒压源 U 正负极，U 调至 10V 左右。R 为 200Ω 限流电阻，防止因 N_1电阻较小造成恒压源输出电流过大。N_2端子 3、4 接入直流毫安表正负端。将恒压源开关闭合和断开，观察并定性记录毫安表读数瞬时正、负的变化，来判定 N_1和 N_2两线圈的同名端。

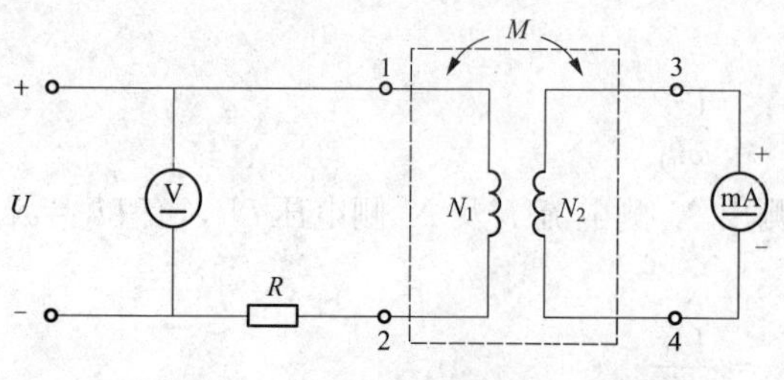

图 14-3　直流法判别同名端电路

（2）交流法。本方法中，由于加在 N_1上的电压仅数伏，直接用实验台内调

压器很难调节，因此采用图 14－4 的线路来增加电压的调节精度。其中 W（或 U、V）、N 为实验台上的自耦调压器的输出端，B 为实验台上铁芯变压器，此处作降压用，即变压器 220V 侧连接自耦调压器输出，36V 侧连接电感线圈。注意 220V/36V 为变比，并非实验中需要的电压。

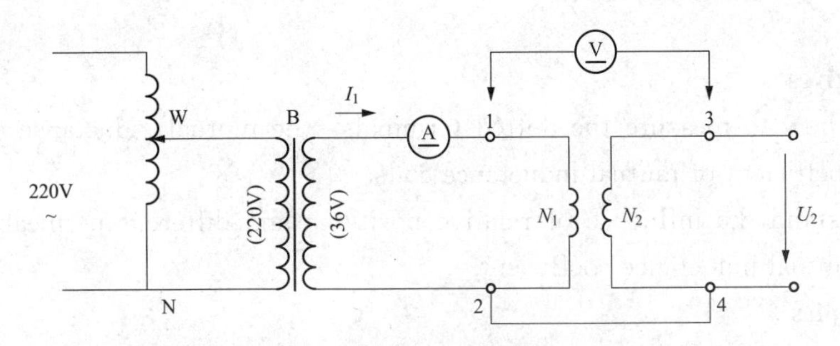

图 14－4　交流法判别同名端电路

接通电源前，应首先检查自耦调压器是否调至零位，确认后方可接通交流电源，缓慢调节调压器手柄，观察电流表，使流过电流表的电流不超过 0.5A，在此前提下自行选择输出电压值，用交流电压表测量 U_{12}、U_{34}、U_{13}，记录数据并判定同名端。

拆去 2、4 连线，将 2、3 相接，再次测量两互感线圈端电压及两线圈未相连的一组端子间的电压，记录数据并判定同名端。

2. 测定两线圈的互感系数 M 和耦合系数 k

在图 14－4 电路中，拆去两线圈间连线，线圈 N_1 侧加电压 U_1，线圈 N_2 开路，控制流过 N_1 侧的电流 I_1 不超过 0.5A，测量并记录 U_1、I_1、U_2；然后在 N_2 侧加电压 U_2，N_1 侧开路，同样控制流过 N_2 侧的电流 I_2 不超过 0.5A，测量并记录 U_2、I_2、U_1。根据以上数据，计算 M_{21} 和 M_{12}。

除去互感线圈连接的电源和接线，用万用表测量并记录两线圈电阻，计算 k。

五、注意事项

（1）整个实验过程中，交流电源和直流电源不得同时连接到电路中。

（2）在直流法判别同名端实验电路中，必须接入限流电阻。

（3）实验前，首先检查变压器 36V 侧的 1.5A 保险丝及互感线圈上的 0.5A 保险丝是否完好。实验中，若电流突然消失，应马上检查保险丝是否熔断。

六、思考题

若本实验使用带有独立铁芯的外接线圈进行，用直流法判断同名端时，可否用插、拔铁芯时观察电流表的正、负读数变化来确定？这与实验原理中所叙述的方法是否一致？

七、实验报告

根据实验内容，总结测定互感线圈同名端的方法。

Experiment 14 Study of Mutual Inductance Circuit

● Objectives

1. Learn how to measure the dotted terminals, the mutual inductance coefficient and the coupling coefficient of mutual inductance coils.

2. Understand the influence of relative positions and different permeability magnetic materials on mutual inductance coefficient.

● Principles

The phenomenon that induction electromotive force is generated in one coil because of the current changing in another coil is called mutual inductance phenomenon, and these two coils are called mutual inductance coils. The performance of mutual inductance coils are judged by mutual inductance coefficient (abbreviated as mutual inductance) M. The geometric sizes, shapes, turn numbers, magnetic conductivity of magnetic materials and the relative positions of the two coils all have influence on mutual inductance.

1. Methods of Determining the Dotted Terminals of Mutual Inductance Coils

(1) DC Method

As shown in Figure 14 - 1, at the moment the switch S is closed, if the pointer of the ammeter deflects positively, then the terminals 1 and 3 can be determined as dotted terminals. If the pointer deflects negatively, the terminals 1 and 4 can be determined as dotted terminals.

(2) AC Method

As shown in Figure 14 - 2, connect a pair of random terminals (e. g. 2 and 4) of the coil N_1 and N_2 together, and add a low voltage to one coil (e. g. N_1) and leave another coil opening. Measure the terminal voltages U_{12} and U_{34} of the two coils and the voltage U_{13} between the two unconnected terminals of the two coils. If U_{13} is the difference of terminal voltages of the two coils, then 1 and 3 are dotted terminal. If U_{13} is the sum of terminal voltages of the two coils, then 1 and 4 are dotted terminal.

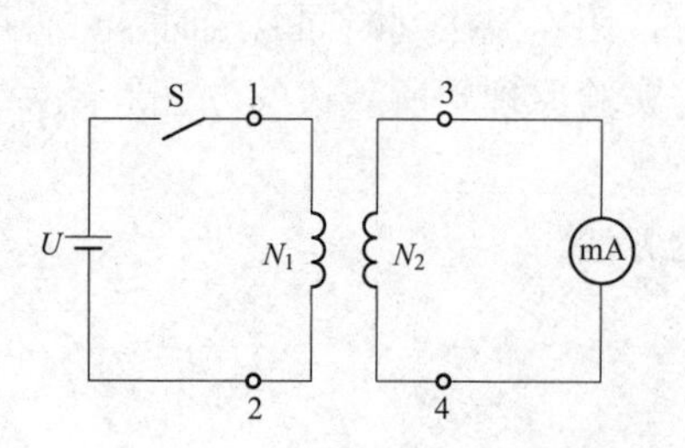

Figure 14 - 1 Principle of DC Method

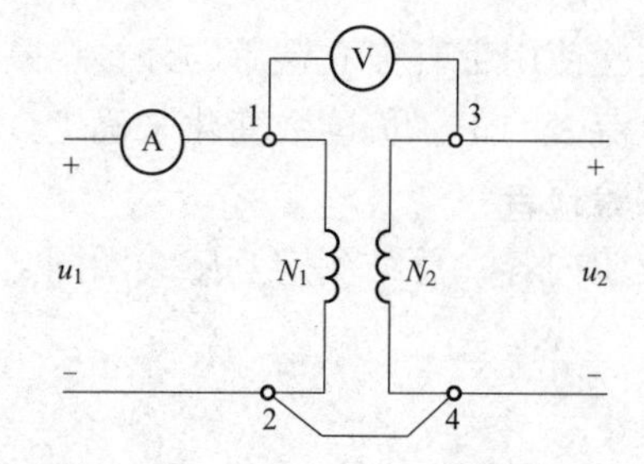

Figure 14 - 2 Principle of AC Method

2. Measurement of Mutual Inductance Coefficient M

In the circuit of Figure 14 - 2, remove the connection between the coils, and add a low

AC voltage U_1 to N_1. Measure the current I_1 through N_1 and the voltage U_2 across N_2. There is no current through N_2 because N_2 is open circuit at this time, so there is $U_2 = \omega MI_1$. The mutual inductance coefficient can be calculated as

$$M_{21} = \frac{U_2}{\omega I_1}$$

Add a low AC voltage U_2 to N_2 and leave N_1 as open circuit. Measure the current I_2 through N_2 and the voltage U_1 across N_1, there is $U_1 = \omega MI_2$. The mutual inductance coefficient can be calculated as

$$M_{12} = \frac{U_1}{\omega I_2}$$

There shall be $M_{21} = M_{12}$, that is, the mutual inductance coefficient M.

3. Measurement of Coupling Coefficient *k*

The degree of coupling of two mutual inductance coils can be expressed by the coupling coefficient k

$$k = M/\sqrt{L_1 L_2}$$

L_1 is the self-inductance coefficient of N_1, and L_2 is the self-inductance coefficient of N_2. The measurement of them is as follows. Add a low AC voltage U_1 to N_1 and leave N_2 open circuit, and measure the current I_1 through N_1. Then add an AC voltage U_2 to N_2 and leave N_1 open circuit, measure the current I_2 through N_2.

When a voltage is added to N_1: $\dot{U}_1 = (R_1 + j\omega L_1)\dot{I}_1 \pm j\omega M\dot{I}_2$.

When a voltage is added to N_2: $\dot{U}_2 = (R_2 + j\omega L_2)\dot{I}_2 \pm j\omega M\dot{I}_1$.

Because N_2 is open circuit when a voltage is added to N_1, and N_1 is open circuit when a voltage is added to N_2.

When a voltage is added to N_1: $\dot{U}_1 = (R_1 + j\omega L_1)\dot{I}_1$.

When a voltage is added to N_2: $\dot{U}_2 = (R_2 + j\omega L_2)\dot{I}_2$.

R_1 and R_2 are the resistances of the mutual inductance coil N_1 and N_2. These resistances can be measured directly on the unconnected N_1 and N_2 by a multimeter, and the self-inductances L_1 and L_2 can be calculated. When the mutual inductance coefficient M is known, k can be calculated.

● Equipment

Equipment is shown in Table 14 - 1.

Table 14 - 1 Equipment

Equipment	Model or Specification	Quantity	Module
Constant Voltage Source	0~30V	1	NDG - 02
DC Voltmeter	0~200V	1	NDG - 03
DC Ammeter	0~2000mA	1	
AC Power Supply	0~450V Three-Phase 0~250V Single-Phase	1	QS - DYD3

续表

Equipment	Model or Specification	Quantity	Module
Smart Meter	0～500V 0～3A	3	NDG-01
Mutual Inductance Coil	30V/0.5A	One Pair	NDG-09
Transformer	220V/36V	1	
Resistor	200Ω	1	NDG-13

● Contents

1. Determination of the Dotted Terminals of Mutual Inductance Coils

(1) DC Method

Connect the circuit according to Figure 14-3. The DC equipment in the circuit shall be connected to the coil through the current sampling sockets to ensure the DC wire and the AC wire are both connected to the appropriate sockets. Connect the terminals 1 and 2 of N_1 to the positive and the negative pole of the DC constant voltage source respectively. Adjust the output voltage U to about 10V. R is a current limiting resistor of 200Ω to prevent excessive output current of the constant voltage source due to the small resistance of N_1. Connect the terminals 3 and 4 of N_2 to the positive and negative terminal of DC ammeter respectively. Turn the constant voltage source on and off, observe and write down the instantaneously positive or negative change of the reading of the ammeter qualitatively. Determine the dotted terminal of N_1 and N_2.

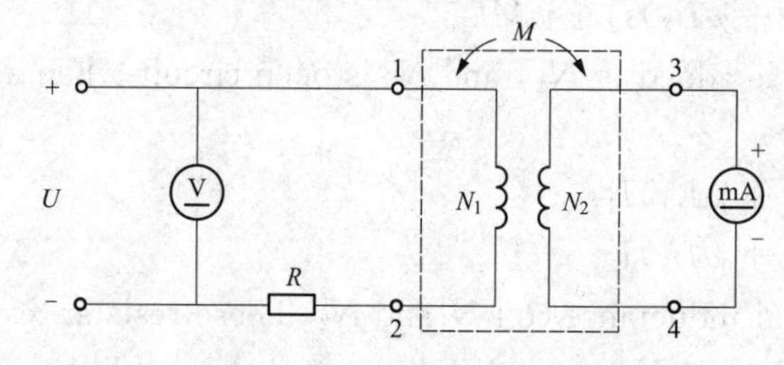

Figure 14-3 Circuit of DC Method

(2) AC Method

In this method, the voltage added to N_1 is only a few volts and hard to adjust with the voltage regulator, thus the circuit in Figure 14-4 is used to increase the voltage regulation precision. W (or U, V) and N are the output terminals of automatic coupling voltage regulator in the experiment station, and B is the iron-core transformer in the experiment station. The transformer is used here for reduction of the voltage, so the 220V side of transformer is connected to the output of automatic coupling voltage regulator and the 36V side is connected to the mutual inductance coils. Please note that 220V/36V is the transformation ratio, not the voltages needed in the experiment.

Before the power is on, check whether the automatic coupling voltage regulator is adjusted to zero output. Turn AC power supply on after the confirmation, and rotate the voltage regulator handle slowly while watching the ammeter to keep the current through the ammeter under 0.5A. Choose an output voltage yourself under this premise, and measure U_{12}, U_{34} and U_{13}. Write down the data and determine the dotted terminals.

Remove the wire between the terminals 2 and 4, and connect the terminals 2 and

3. Measure the terminal voltages across the two mutual inductance coils and the voltage between the two unconnected terminals of the two coils. Write down the data and determine the dotted terminals.

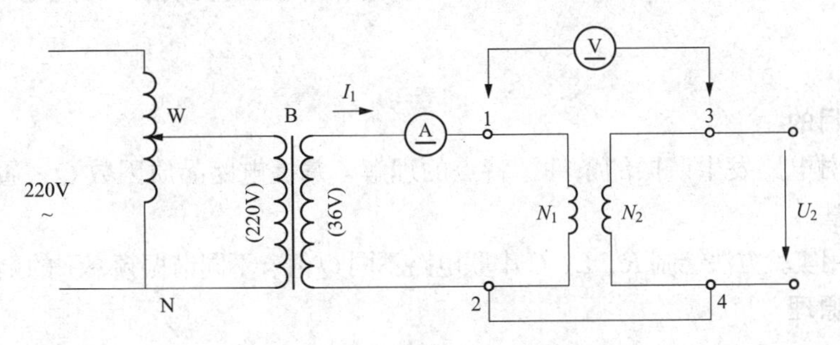

Figure 14 - 4 Circuit of AC Method

2. Measurement of the Mutual Inductance Coefficient M and the Coupling Coefficient k

Remove the wire between the two coils in the circuit of Figure 14 - 4. Add a voltage U_1 to coil N_1 and leave coil N_2 open circuit. Adjust the voltage and keep the current I_1 through N_1 under 0. 5A. Measure and write down U_1, I_1 and U_2. Then add a voltage U_2 to coil N_2 and leave coil N_1 open circuit. Adjust the voltage and keep the current I_2 through N_2 under 0. 5A. Measure and write down U_2, I_2 and U_1. Calculate M_{21} and M_{12} using the above data.

Remove the power supply and the wire connected to the mutual inductance coils. Measure the resistances of the two coils with a multimeter, and calculate k.

● **Notes**

1. In the whole process of experiment, AC and DC power supply must not be connected to the circuit at the same time.

2. A current limiting resistor must be connected into the circuit of determining the dotted terminal by the DC method.

3. Check whether the 1. 5A fuse at the 36V side of the transformer and the 0. 5A fuse at the mutual inductance coil are in good conditions. In the process of the experiment, if the current vanishes, check if the fuses have fused immediately.

● **Questions**

Can the method of observing the positive and negative reading of the ammeter while inserting and pulling out the iron core be used to determine the dotted terminal in the DC method step if some external mutual inductance coils with independent iron cores are used in this experiment? Is it equivalent with the DC method mentioned in the principle part?

● **Experiment Report**

Summarize the methods of measuring the dotted terminal of mutual inductance coils using the experiment contents.

实验 15　R、L、C 串联谐振电路的研究

一、实验目的

（1）加深对电路发生谐振的条件、特点的理解，掌握电路品质因数 Q、通频带的物理意义及其测定方法。

（2）学习用实验方法绘制 R、L、C 串联电路不同 Q 值、不同谐振频率下的幅频特性曲线。

二、实验原理

在图 15-1 所示的 R、L、C 串联电路中，电路复阻抗 $Z=R+\mathrm{j}\left(\omega L-\dfrac{1}{\omega C}\right)$，当 $\omega L=\dfrac{1}{\omega C}$ 时，$Z=R$，$\dot{U}$ 与 $\dot{I}$ 同相，电路发生串联谐振，谐振角频率 $\omega_0=\dfrac{1}{\sqrt{LC}}$，谐振频率 $f_0=\dfrac{1}{2\pi\sqrt{LC}}$。

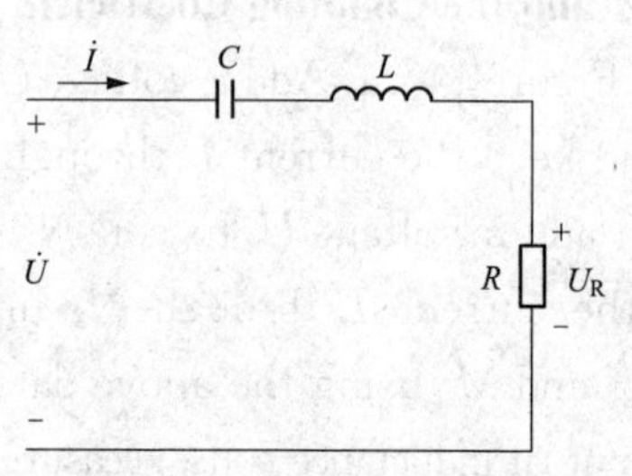

图 15-1　RLC 串联电路

在图 15-1 电路中，若 $\dot{U}$ 为激励信号，$\dot{U}_R$ 为响应信号，其幅频特性曲线如图 15-2 所示，在 $f=f_0$ 时，$A=1$，$U_R=U$；$f\neq f_0$ 时，$U_R<U$，电路呈带通特性。$A=0.707$，即 $U_R=0.707U$ 所对应的两个频率 f_L 和 f_H 为下限频率和上限频率，f_H-f_L 为通频带。通频带的宽窄与电阻 R 有关，不同电阻值的幅频特性曲线如图 15-3 所示。

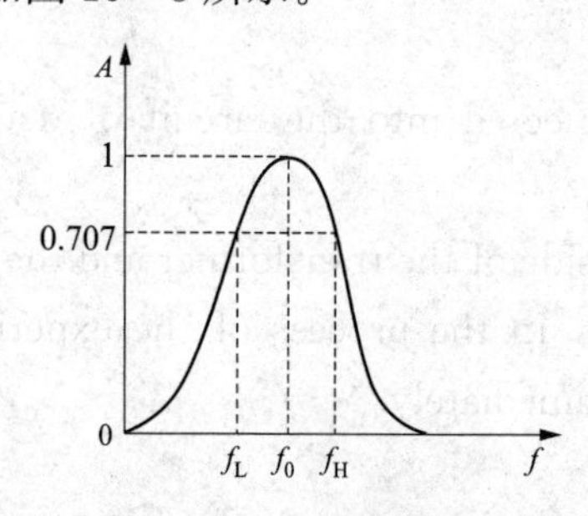

图 15-2　U_R 的幅频特性曲线

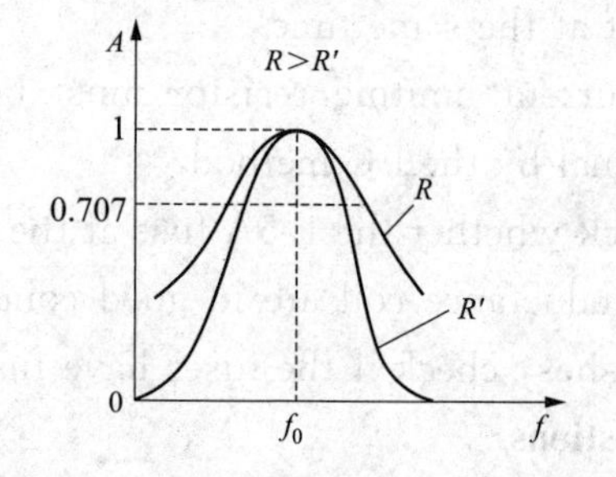

图 15-3　不同电阻值下的幅频特性曲线

电路发生串联谐振时，$U_R=U$，$U_L=U_C=QU$，Q 称为品质因数，有

$$Q=\frac{U_L(\omega_o)}{U}=\frac{U_C(\omega_o)}{U}=\frac{\omega_o L}{R}=\frac{1}{\omega_o CR}=\frac{1}{R}\sqrt{L/C}$$

可知 Q 与电路的参数 R、L、C 有关。Q 值越大，幅频特性曲线越尖锐，通频带越窄，电路的选择性越好。在恒压源供电时，电路的品质因数、选择性与通频带只决定于电路本身的参数，而与信号源无关。

在本实验中，测量输入信号不同频率下的电压 U_R、U_L、U_C，绘制 R、L、C 串联电路的幅频特性曲线，并根据 $\Delta f=f_H-f_L$ 计算出通频带，根据 $Q=\dfrac{U_L}{U}=\dfrac{U_C}{U}$ 或 $Q=\dfrac{f_0}{f_H-f_L}$ 计算出品质因数。

三、实验设备

实验设备见表 15-1。

表 15-1　**实　验　设　备**

设备名称	型号与规格	数量	实验模块
信号发生器	DG1022U	1	
交流毫伏表	10Hz～1MHz 100μV～700V	1	NDG-05
万用表	UT803	1	
电阻、电感、电容	多种		NDG-13
实验电路	串联谐振电路	1	NDG-12

四、实验内容

(1) 在电路模块 NDG-13 上选取元件 $L=10\text{mH}$，$R=51\Omega$，$C=0.033\mu\text{F}$，按图 15-4 组成串联电路。信号源输出有效值为 3V 的正弦波信号（用毫伏表或万用表测量）作为电路激励并保持幅值不变。也可使用 NDG-12 上的串联谐振电路（$L=9\text{mH}$，$R=51\Omega$，$C=0.033\mu\text{F}$）。

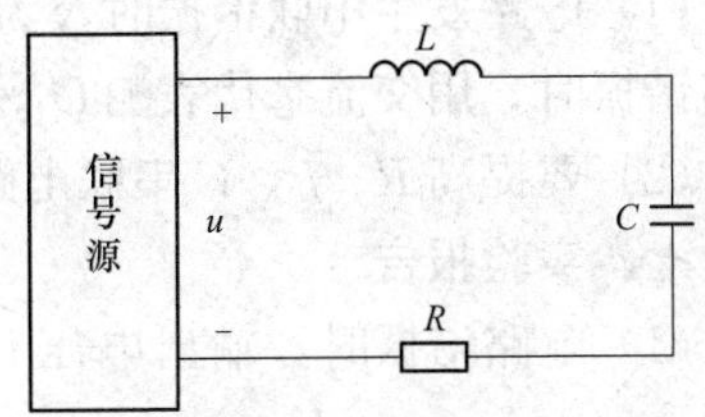

图 15-4　RLC 串联实验电路

(2) 测量 R、L、C 串联电路谐振频率。调节信号源正弦波输出信号频率，由小逐渐变大，并用毫伏表或万用表测量电阻 R 两端电压 U_R，当 U_R 的数值为最大时，激励信号频率即为电路的谐振频率 f_0，测量此时的 U_C 与 U_L 值，将测量数据记入表 15-2 中。

(3) 测量 R、L、C 串联电路的幅频特性。调节信号源正弦波输出信号频率，在上述实验电路的谐振点两侧，自行选择频率步进值，依次各取 7 个测量点，注意此 14 个点中必须包括 f_L 和 f_H，逐点测出 U_R、U_L 和 U_C 值，记入表 15-2 中。

表 15-2　**R、L、C 串联电路的幅频特性数据 1**

f (kHz)														
U_R (V)														
U_L (V)														
U_C (V)														
$f_0=$						$f_H-f_L=$						$Q=$		

(4) 在自行连接的电路中，保持 L、C 不变，选择其他电阻，组成串联电路。或使用 NDG-12 上 $R=100\Omega$ 的串联谐振电路。重复步骤 2、3 的测量过程，将幅频特性数据记入表 15-3 中。

表 15-3　**R、L、C 串联电路的幅频特性数据 2**

f (kHz)														
U_R (V)														
U_L (V)														
U_C (V)														
$f_0=$						$f_H-f_L=$						$Q=$		

(5) 在自行连接的电路中，$R=51\Omega$，L、C 其中一个元件保持不变，另一元件选择其他

参数，组成串联电路（不使用NDG－12）。重复步骤2、3的测量过程，将幅频特性数据记入表15－4中。

表15－4　　R、L、C串联电路的幅频特性数据3

f（kHz）															
U_R（V）															
U_L（V）															
U_C（V）															
f_0＝						f_H-f_L＝						Q＝			

五、注意事项

测试频率点的选择应在靠近谐振频率附近多取几点。在改变频率时，应注意信号输出电压，使其维持在3V不变。

六、思考题

（1）电路发生串联谐振时，为什么输入电压u不能太大？如果信号源给出3V的电压，电路谐振时，用交流毫伏表测U_L和U_C，应该选择用多大的量限？为什么？

（2）要提高R、L、C串联电路的品质因数，电路参数应如何改变？

七、实验报告

（1）电路谐振时，输出电压U_R与输入电压U是否相等？U_L和U_C是否相等？试分析原因。

（2）根据表15－2～表15－4测量数据，分别绘出三条幅频特性曲线$U_R=f_R(f)$、$U_L=f_L(f)$、$U_C=f_C(f)$。

（3）计算出通频带与Q值，说明不同R值时对电路通频带与品质因数的影响。

（4）对两种不同的测Q值的方法进行比较，分析误差原因。

（5）试总结串联谐振的特点。

Experiment 15 Study of Series R, L, C Resonant Circuit

● **Objectives**

1. Acquire a better understanding of conditions and characteristics of resonance in a circuit, and understand the physical meaning and measurement of quality factor Q and passband.

2. Learn how to draw the amplitude frequency characteristic curves under different Q values and different resonant frequencies of series R, L, C resonant circuit by experimental method.

● **Principles**

In the series R, L, C resonant circuit shown in Figure 15 - 1, the complex impedance $Z = R + \mathrm{j}\left(\omega L - \frac{1}{\omega C}\right)$. When $\omega L = \frac{1}{\omega C}$, we get $Z = R$, at this time $\dot{U}$ and $\dot{I}$ are the same phase, series resonance occurred in the circuit. The resonant angle frequency $\omega_0 = \frac{1}{\sqrt{LC}}$, and the resonant frequency $f_0 = \frac{1}{2\pi\sqrt{LC}}$.

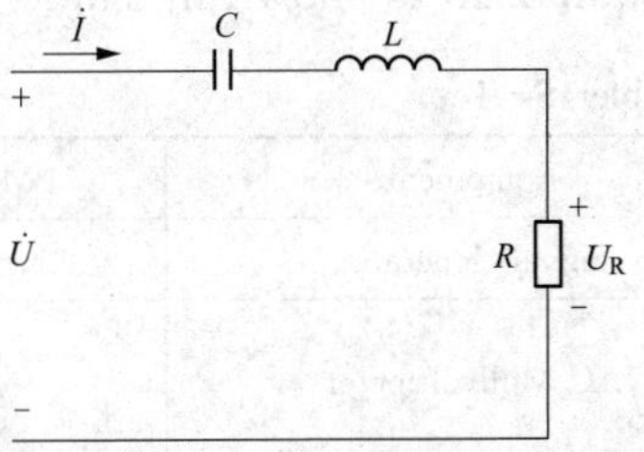

Figure 15 - 1 The Series RLC Circuit

In the circuit of Figure 15 - 1, if U is the excitation signal and U_R is the response signal, the amplitude frequency characteristic curve is shown in Figure 15 - 2. At the point $f = f_0$, we get $A = 1$, $U_R = U$. When $f \neq f_0$, $U_R < U$, the circuit shows band-pass characteristics. The two frequencies f_L and f_H corresponding to $A = 0.707$, namely, $U_R = 0.707U$, are upper limit frequency and lower limit frequency. $f_H - f_L$ is the passband. The width of the passband is related to the resistance R. The amplitude frequency characteristic curves corresponding to difference resistance are shown in Figure 15 - 3.

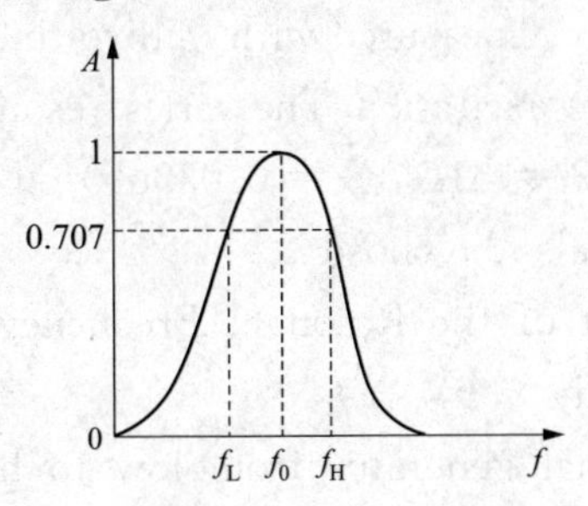

Figure 15 - 2 The Amplitude Frequency Characteristic Curve of U_R

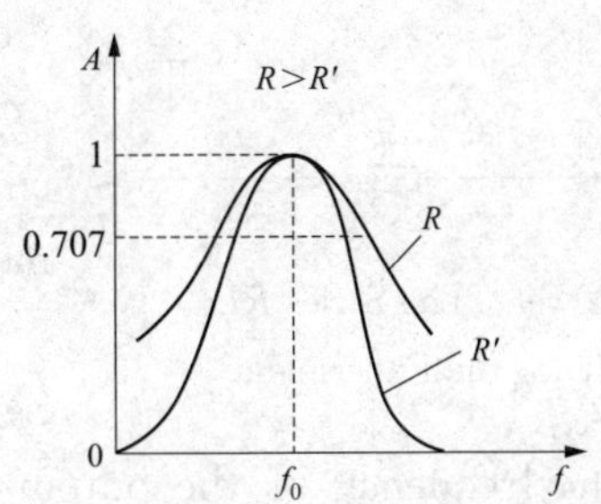

Figure 15 - 3 The Amplitude Frequency Characteristic Curves Corresponding to Difference Resistances

When series resonance occurred in the circuit, $U_R = U$, $U_L = U_C = QU$, Q is called quality factor, there is

$$Q=\frac{U_L(\omega_o)}{U}=\frac{U_C(\omega_o)}{U}=\frac{\omega_o L}{R}=\frac{1}{\omega_o CR}=\frac{1}{R}\sqrt{L/C}$$

It can be seen that Q is related to the parameters of the circuit R, L, C. The bigger the Q value is, the sharper the frequency characteristic curve is, and the better the selectivity of the circuit is. When the power source is unchanged, the quality factor, the selectivity and the passband are determined only by the parameters of the circuit itself and are unrelated to the signal source.

In this experiment, measure the voltages U_R, U_L, U_C under the input signals at different frequencies, and draw the amplitude frequency characteristic curves of R, L, C series resonant circuit. Calculate the passband according to $\Delta f=f_H-f_L$, and calculate the quality factor according to $Q=\frac{U_L}{U}=\frac{U_C}{U}$ or $Q=\frac{f_0}{f_H-f_L}$.

- **Equipment**

Equipment is shown in Table 15 - 1.

Table 15 - 1 Equipment

Equipment	Model or Specification	Quantity	Module
Signal Generator	DG1022U	1	
AC Millivoltmeter	10Hz～1MHz 100μV～700V	1	NDG - 05
Multimeter	UT803	1	
Resistor, Inductor, Capacitor	Multiple		NDG - 13
Experiment Circuit	Series Resonant circuit	1	NDG - 12

- **Contents**

1. Choose the elements $L=10\text{mH}$, $R=51\Omega$, $C=0.033\text{uF}$ on the module NDG-13. Connect the series circuit in Figure 15 - 4. The signal generator outputs a sinusoidal wave signal with a constant RMS of 3V (measured with millivoltmeter or multimeter) as the excitation. The series resonant circuit ($L=9\text{mH}$, $R=51\Omega$, $C=0.033\text{uF}$) on the module NDG-12 can also be chosen.

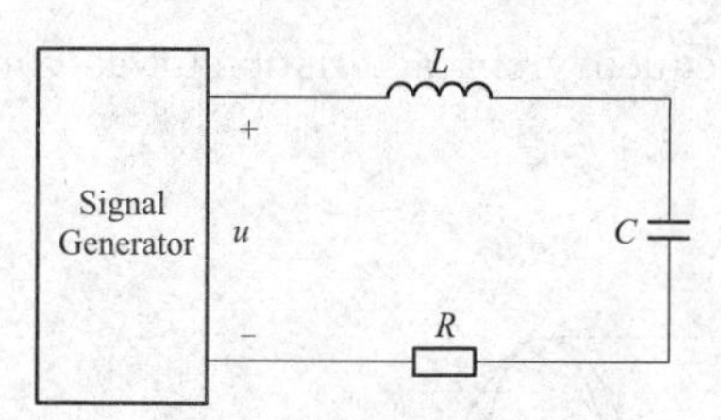

Figure 15 - 4 The Series *RLC* Circuit for the Experiment

2. Measurement of the Resonant Frequency of Series R, L, C Circuit

Adjust the frequency of the output signal of the signal generator from low to high gradually, and measure the voltage U_R across the resistor R with a millivoltmeter or a multimeter. The frequency of the excitation signal is the resonant frequency f_0 when the value of U_R is maximum. Measure U_C and U_L at this moment, and fill the data into Table

15 - 2.

3. Measurement of the Amplitude Frequency Characteristic of Series R, L, C Circuit

Adjust the frequency of the output signal of the signal generator, and choose seven measurement points on both sides of the resonant point by yourself. Please note that f_L and f_H must be included in these 14 points. Measure U_R, U_L, and U_C point by point, and fill in Table 15 - 2.

Table 15 - 2 Data 1 of the Amplitude Frequency Characteristic of Series *R*, *L*, *C* Circuit

f (kHz)															
U_R (V)															
U_L (V)															
U_C (V)															
f_0=					f_H-f_L=						Q=				

4. In the circuit you connected, keep L and C unchanged and choose another resistor to compose a new series circuit. The series resonant circuit (R = 100Ω) on the module NDG-12 can also be chosen. Repeat the measuring process of step 2 and 3, and fill in Table 15 - 3.

Table 15 - 3 Data 2 of the Amplitude Frequency Characteristic of Series *R*, *L*, *C* Circuit

f (kHz)															
U_R (V)															
U_L (V)															
U_C (V)															
f_0=					f_H-f_L=						Q=				

5. In the circuit you connected, R = 51Ω, keep one of the L, C elements unchanged and change the parameter of another to compose a new series circuit (NDG-12 is not used in this step). Repeat the measuring process of step 2 and 3, and fill in Table 15 - 4.

Table 15 - 4 Data 3 of the Amplitude Frequency Characteristic of Series *R*, *L*, *C* Circuit

f (kHz)															
U_R (V)															
U_L (V)															
U_C (V)															
f_0=					f_H-f_L=						Q=				

- **Notes**

The frequency measurement point should be chosen a little densely near the resonant frequency. The output voltage should be keep at 3V constantly when the frequency is changed.

- **Questions**

1. Why can't the input voltage u be too high when series resonance occurred in the

circuit? What range of AC millivoltmeter should be chosen to measure U_L and U_C when resonance occurred in the circuit if the output voltage of the signal generator is 3V? Why?

2. How to change the parameters of the circuit to improve the quality factor of series R, L, C circuit?

- **Experiment Report**

1. Is the output voltage U_R equal to the input voltage U when resonance occurred in the circuit? Is U_L equal to U_C? Try to analyze the reason.

2. Using the data of Table 15－2 ～Table 15－4, draw the three amplitude frequency characteristic curves $U_R = f_R(f)$, $U_L = f_L(f)$, $U_C = f_C(f)$ respectively.

3. Calculate the passbands and values of Q, and explain the influence of different values of R on the passband and the quality factor.

4. Compare the two different methods of measuring the value of Q, and analyze the cause of error.

5. Try to summarize the characteristics of series resonance.

实验 16　三相电路电压、电流与有功功率的测量

一、实验目的

（1）掌握三相负载星形连接、三角形连接的方法。

（2）了解三相电路线电压与相电压，线电流与相电流之间的关系。

（3）了解三相四线制供电系统中中线的作用。

（4）观察线路故障时的情况。

（5）学会用功率表测量三相电路功率的方法。

二、实验原理

（1）在三相电路中，电源用三相四线制向负载供电。三相负载可接成星形（又称 Y 形）或三角形（又称△形）。

当三相对称负载 Y 形连接时，线电压 U_L 是相电压 U_P 的 $\sqrt{3}$ 倍，线电流 I_L 等于相电流 I_P，即

$$U_L = \sqrt{3}U_P, I_L = I_P$$

在理想情况下，流过中线的电流 $I_0=0$，所以可以省去中线。

在对称三相负载作△形联结时，线电压 U_L 等于相电压 U_P，线电流 I_L 是相电流 I_P 的 $\sqrt{3}$ 倍，即

$$I_L = \sqrt{3}U_P, U_L = U_P$$

不对称三相负载 Y 形连接时，必须采用 Y_0 形接法，即三相四线制接法。中线必须牢固连接，以保证三相不对称负载的每相电压等于电源的对称相电压。若中线断开，会导致三相负载电压的不对称，致使负载轻的那一相的相电压过高，使负载遭受损坏；负载重的一相相电压又过低，使负载不能正常工作。

对于不对称负载△连接时，$I_L \neq \sqrt{3}\, I_P$，但只要电源的线电压 U_L 对称，加在三相负载上的电压仍是对称的，对各相负载工作没有影响。

本实验中，用三相调压器调压输出作为三相交流电源，用三组白炽灯作为三相负载，线电流、相电流、中线电流用电流插头和插座测量。

（2）对于三相四线制供电的三相 Y_0 连接的负载，可用三只功率表（或一只功率表）测量各相的有功功率 P_A、P_B、P_C，则三相功率之和 $\sum P = P_A + P_B + P_C$ 即为三相负载的总有功功率值，这就是三瓦特表法（一瓦特表法），如图 16－1 所示。线路中的电流表和电压表用以监视流过功率表的电流和功率表两端电压，使之不超过功率表电流、电压量程。若三相负载是对称的，则只需测量一相的功率，再乘以 3 即得三相总的有功功率。

（3）三相三线制供电系统中，不论三相负载是否对称，也不论负载是 Y 接还是△接，都可用二瓦特表法测量三相负载的有功功率。二瓦特表接法如图 16－2 所示，若两个功率表的读数为 P_1、P_2，则三相功率为

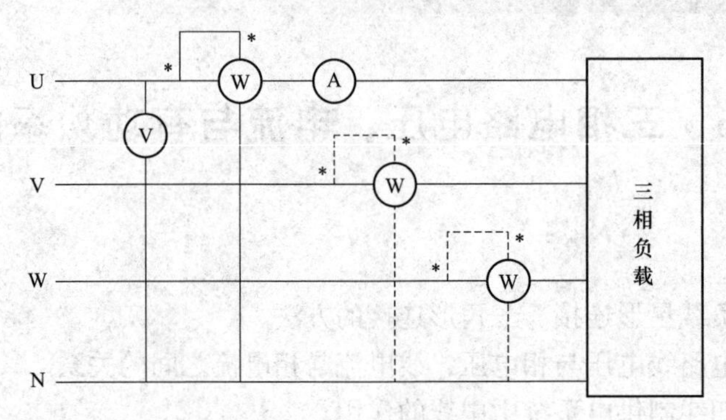

图 16-1 三瓦特表法

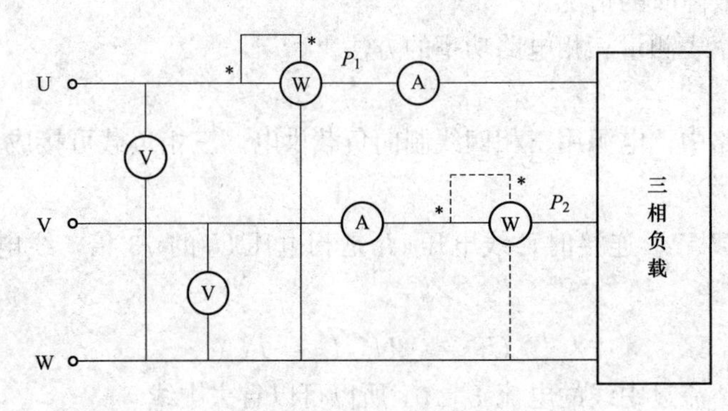

图 16-2 二瓦特表法

$$P = P_1 + P_2 = U_{UW} I_U \cos(30° - \varphi) + U_{VW} I_V \cos(30° + \varphi)$$

式中：φ 为负载的阻抗角（即功率因数角）。

两个功率表的读数与 φ 有下列关系：

1）当负载为纯电阻且对称，$\varphi=0$，$P_1 = P_2$，即两个功率表读数相等。

2）当负载功率因数 $\cos\varphi=0.5$，$\varphi=\pm 60°$，将有一个功率表的读数为零。

3）当负载功率因数 $\cos\varphi<0.5$，$|\varphi|>60°$，则有一个功率表的读数为负值。

三、实验设备

实验设备见表 16-1。

表 16-1 实验设备

设备名称	型号与规格	数量	实验模块
智能仪表	0～500V 0～3A	3	NDG-01
交流电源	0～450V 三相/0～250V 单相		QS-DYD3
白炽灯泡	25W	9	NDG-10
电容	2.2μF/500V	1	NDG-08

四、实验内容

1. 三相负载 Y_0/Y 形连接电压和电流的测量

实验电路以实物为准，将白炽灯参考图 16-3 所示连接成 Y 形接法，用三相调压器调压输出作为三相交流电源。连接好线路前，必须将三相调压器的手柄旋至零位（逆时针旋到底）。接好线路后缓慢调节调压器的输出，使输出的三相线电压为 220V。按以下要求测量并记录数据。

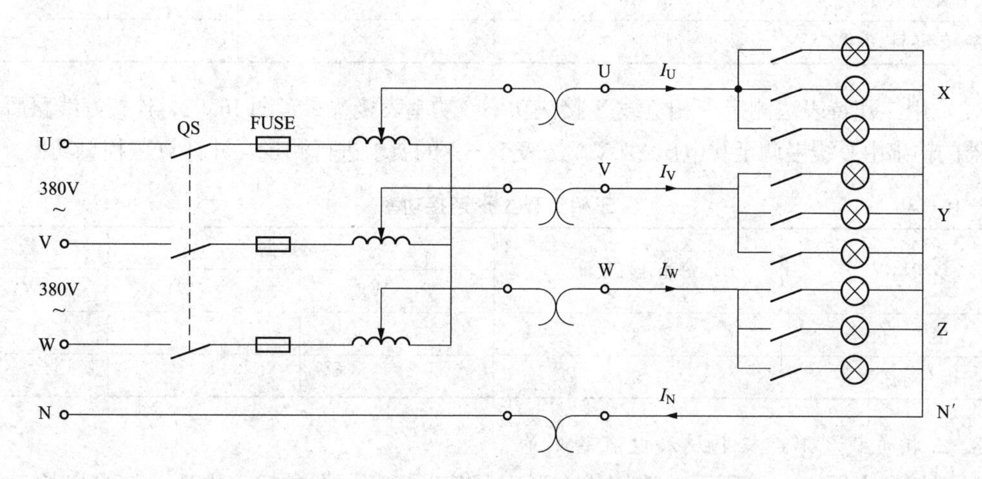

图 16－3　三相负载 Y_0/Y 形连接

（1）连接 N 与 N′在三相负载有中线的情况下（Y_0接），测量负载对称、不对称、一相负载断开时的各相电流、中线电流（利用电流取样插座）和各相负载电压，将数据记入表 16－2 中，并比较各相白炽灯亮度。

（2）去掉 N 与 N′间连线在三相负载无中线的情况下（Y 接），测量负载对称、不对称、一相负载断开时的各相电流、各相电压和三相电源中点 N 到负载中点 N′的电压（简称中点电压 $U_{NN'}$），将数据记入表 16－2 中，并比较各相白炽灯亮度。

表 16－2　三相负载 Y_0/Y 形连接电压、电流

中线连接	每相开灯数量	负载相电压（V）			电流（A）				$U_{NN'}$（V）	亮度比较 U、V、W
		$U_{UN'}$	$U_{VN'}$	$U_{WN'}$	I_U	I_V	I_W	I_N		
有（Y_0）	3、3、3								—	
	1、2、3								—	
	3、0、3								—	
无（Y）	3、3、3							—		
	1、2、3							—		
	3、0、3							—		

2. 三相负载 Y_0/Y 形联结功率的测量

（1）用一瓦特表法测定三相负载 Y_0接法功率。功率表接法参考图 16－1。连接好线路后，将调压器的输出由 0 缓慢调至线电压 220V，按表 16－3 的要求进行测量，并计算三相总功率。

表 16-3 三相负载 Y_0 形连接功率

负载情况	每相开灯数量	测量数据			计算值
		P_U（W）	P_V（W）	P_W（W）	ΣP（W）
Y_0接对称负载	3、3、3				
Y_0接不对称负载	1、2、3				

（2）用二瓦特表法测定三相负载 Y 接法功率。功率表接法参考图 16-2，连接好线路后，将调压器的输出由 0 缓慢调至线电压 220V，按表 16-4 的要求进行测量，并计算三相总功率。

表 16-4 三相负载 Y 形连接功率

负载情况	每相开灯数量	测量数据		计算值
		P_1（W）	P_2（W）	ΣP（W）
Y 接对称负载	3、3、3			
Y 接不对称负载	1、2、3			

3. 三相负载△形连接电压和电流的测量

白炽灯参考图 16-4 所示，连接成△形。调节三相调压器的输出电压，使输出的三相线电压为 150V。测量三相负载对称和不对称时的各相电流、线电流（利用电流取样插座）和三相负载电压，将数据记入表 16-5 中，并比较各相白炽灯亮度。

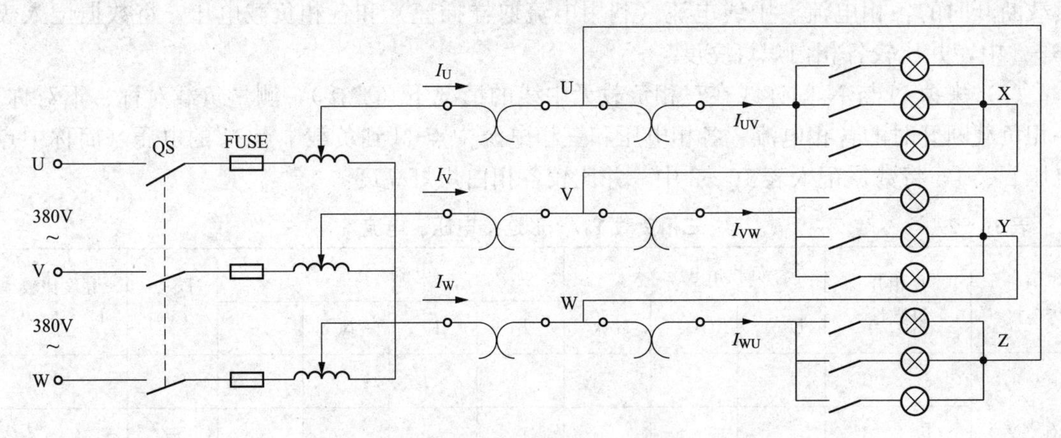

图 16-4 三相负载△形连接

表 16-5 三相负载△形连接电压、电流

每相开灯数量	负载电压（V）			线电流（A）			相电流（A）			亮度比较
	U_{UV}	U_{VW}	U_{WU}	I_U	I_W	I_V	I_{UV}	I_{VW}	I_{WU}	
3、3、3										
1、2、3										

4. 三相负载△形连接功率的测量

用二瓦特表法测定三相三线制△接法电路功率。功率表接法如图 16-3 所示，连接好线路后，将调压器的输出由 0 缓慢调至线电压 150V，按表 16-6 的要求进行测量，并计算三相总功率。

表 16-6　**三相负载△形连接功率**

负载情况	每相开灯数量	测量数据		计算值
		P_1（W）	P_2（W）	ΣP（W）
△接对称负载	3、3、3			
△接不对称负载	1、2、3			

五、注意事项

（1）实验之前，以及每次改变接线时，均必须将三相调压器调压手柄调至零位。

（2）必须严格遵守实验开始时先接线，后通电；实验结束时先断电，后拆线的实验操作原则。

（3）在测量△形负载线电流和相电流时，应注意测量位置的区别。

（4）加电压前应先打开灯泡开关，并在加压过程中监视电源电压大小。如电源电压超过100V灯泡仍未点亮，应将调压器回零并检查原因，不可持续加压以至灯泡电压超过额定电压。

六、思考题

（1）本实验Y形电路中为什么将三相电源线电压设定为220V？

（2）中线上能安装保险丝吗？为什么？

（3）测量功率时为什么在线路中通常接有电流表和电压表？

七、实验报告

（1）根据实验数据，在负载为Y形连接时，$U_L=\sqrt{3}U_P$ 在什么条件下成立？在△形连接时，$I_L=\sqrt{3}I_P$ 在什么条件下成立？

（2）根据对称和不对称负载Y形连接时的负载相电压值绘出相量图。并根据实验数据和观察到的现象，总结三相四线制供电系统中中线的作用。

（3）不对称△形连接的负载，能否正常工作？实验是否能证明这一点？

（4）根据不对称负载△形连接时的相电流值作相量图，并求出线电流值，然后与实验测得的线电流作比较，进行分析。

（5）比较一瓦特表和二瓦特表法的测量结果，总结分析三相电路功率测量的方法与结果。

Experiment 16　Measurement of Voltage, Current and Active Power of Three-Phase Circuits

● **Objectives**

1. Understand the two basic connections of three-phase loads: Y-connected load and △-connected load.

2. Understand the relationships between the line voltages and the phase voltages, and the relationships between the line current and the phase current of three-phase circuit.

3. Understand the role of the neutral wire in a three-phase four-wire power supply system.

4. Observe the states of line faults.

5. Learn how to measure the active power of a three-phase circuit with power meters.

● **Principles**

1. In a three-phase circuit, the power supply is in a three-phase four-wire system. The three-phase load can be connected as a star shape (aka Y-connected), or a triangle shape (aka △-connected).

When a three-phase symmetrical load is Y-connected, the line voltage U_L is $\sqrt{3}$ times the phase voltage U_P, and the line current I_L is equal to the phase current I_P. There are

$$U_L = \sqrt{3}U_P, I_L = I_P$$

In the ideal situation, the current through the neutral wire $I_0 = 0$, and the neutral wire can be omitted.

When a three-phase symmetrical load is △-connected, the line voltage U_L is equal to the phase voltage U_P, the line current I_L is $\sqrt{3}$ times the phase current I_P. There are

$$I_L = \sqrt{3}U_P, U_L = U_P$$

When a three-phase asymmetrical load is Y-connected, a neutral wire must be used. A three-phase Y-connected connection with a neutral wire is called three-phase four-wire connection. The neutral wire must be connected firmly to guarantee the voltage of each phase of three-phase asymmetrical load is equal to the symmetrical phase voltage of power supply. The disconnection of neutral wire will lead to. The three-phase voltage asymmetry will cause the phase voltage of the phase of light load too high to damage the load and the phase voltage of the phase of heavy load too low to keep the load work normally.

When a three-phase asymmetrical load is △-connected, there is $I_L \neq \sqrt{3} I_P$. But as long as the line voltages of the power supply are symmetrical, the voltages added to the three phase load are symmetrical too.

In this experiment, the output voltage of the three phase voltage regulator is used as a three-phase AC power supply, and three sets of incandescent lamps are used as a three-phase load. The line current, the phase current and the neutral wire current are measured through current plugs and current sampling sockets.

2. To a three-phase Y_0-connected load which is supplied by a three-phase four-wire power supply, three power meters (or only one power meter) can be used to measure the active powers of each phase. The powers of each phase are called P_A, P_B, P_C. The sum of the three powers $\sum P = P_A + P_B + P_C$ is the total power of three-phase load. This is three-wattmeter method (one-wattmeter method), as shown in Figure 16 - 1. The ammeter and the voltmeter are used to monitor the current through the power meter and the voltage across the power meter to not exceed the current and voltage range of the power meter. If the three-phase load is symmetrical, then only one phase power needs to be measured, and the three-phase total active power is obtained by multiplying it by 3.

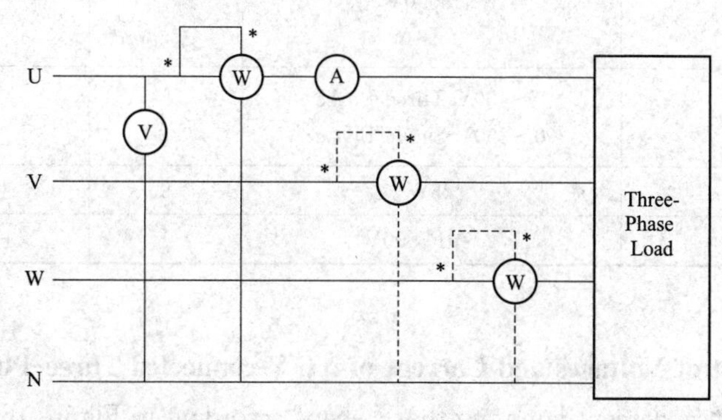

Figure 16 - 1 Three-Wattmeter Method

3. In a three-phase three-wire system, whether the three-phase load is symmetrical or not, and no matter the load is Y-connected or △-connected, the two-wattmeter method can be used to measure the active power of three-phase load universally. The two-wattmeter method is shown in Figure 16 - 2, if the readings of the two power meter are P_1 and P_2, the three-phase power

$$P = P_1 + P_2 = U_{UW} I_U \cos(30° - \varphi) + U_{VW} I_V \cos(30° + \varphi)$$

φ is the impedance angle (power factor angle) of the load.

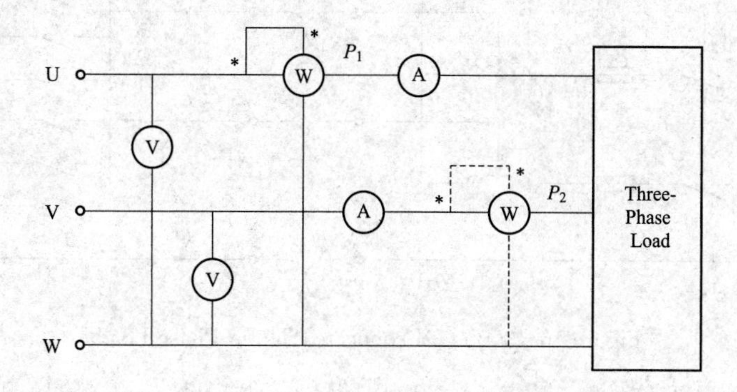

Figure 16 - 2 Two-Wattmeter Method

The relationships of the readings of the two power meters and φ:

(1) The loads is pure resistance and symmetrical, $\varphi = 0$, $P_1 = P_2$, the readings of the two power meters are equal.

(2) The power factor of the load $\cos\varphi = 0.5$, $\varphi = \pm 60°$, there will be a zero reading from one of the power meters.

(3) The power factor of the load $\cos\varphi < 0.5$, $|\varphi| > 60°$, there will be a negative reading from one of the power meters.

- **Equipment**

Equipment is shown in Table 16 - 1.

Table 16 - 1 Equipment

Equipment	Model or Specification	Quantity	Module
Smart Meter	0～500V 0～3A	3	NDG - 01
AC Power supply	0～450V Three-Phase 0～250V Single-Phase		QS-DYD3
Incandescent Bulb	25W	9	NDG - 10
Capacitor	2.2μF/500V	1	NDG - 08

- **Contents**

1. Measurement of Voltages and Current of Y_0/Y-connected Three-Phase Load

Connecting the incandescent lamps to the Y shape according to Figure 16 - 3. The three-phase AC power supply is the output of the three-phase voltage regulator. Before the circuit connection is completed, the handle of the three-phase voltage regulator must be put in zero position. After completing the connection, adjust the output line voltage of the voltage regulator slowly to 220V. Measure and write down the data according to the following requirements.

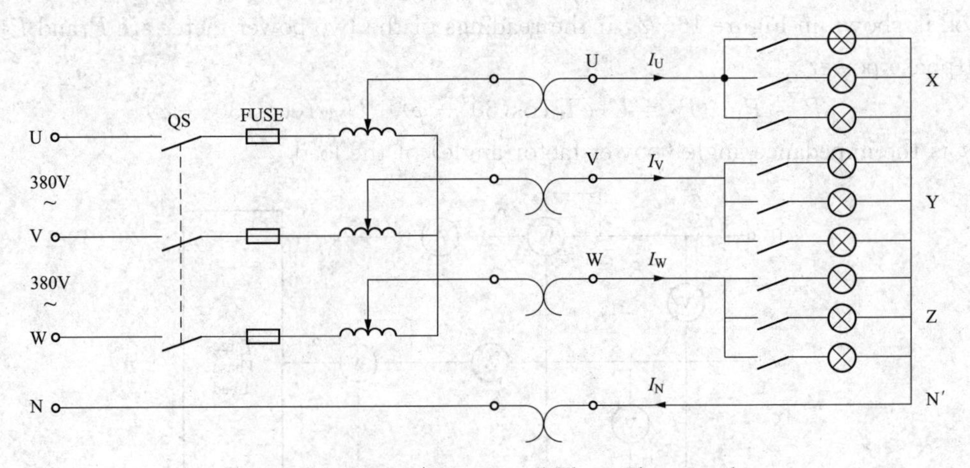

Figure 16 - 3 Y_0/Y-connected Three-Phase Load

(1) In the case of a neutral wire is connected to the three-phase load (Y_0-connected), Connect N and N′ and measure the current of each phase, neutral current (through the current sampling sockets) and the load voltage of each phase when the load is symmetrical

and asymmetrical and lack of one phase. Fill in Table 16 - 2 and compare the brightness of incandescent lamps in each phase.

(2) In the case of no neutral wire is connected to the three-phase load (Y-connected), remove the wire between N and N′ and measure the current of each phase, the load voltage of each phase and the voltage across the neutral point N of the three-phase power supply and the neutral point N′ of the load (neutral point voltage $U_{NN'}$ for short) when the load is symmetrical and asymmetrical and lack of one phase. Fill in Table 16 - 2 and compare the brightness of incandescent lamps in each phase.

Table 16 - 2　The Voltages and the Current of Y_0/Y-connected Three-Phase Load

Neutral Wire	Number Of Lights Per Phase	Load Phase Voltage (V)			Current (A)				$U_{NN'}$ (V)	Brightness Comparison
		$U_{UN'}$	$U_{VN'}$	$U_{WN'}$	I_U	I_V	I_W	I_N		
Yes (Y_0 - connected)	3、3、3								—	
	1、2、3								—	
	3、0、3								—	
No (Y-connected)	3、3、3							—		
	1、2、3							—		
	3、0、3							—		

2. Measurement of the Power of Y_0/Y-connected Three-Phase Load

(1) Measure the power of Y_0-connected load by one-wattmeter method. The connection of the power meter refers to Figure 16 - 1. Adjust the output line voltage of the voltage regulator from 0 to 220V slowly, measure the data according to Table 16 - 3 and calculate the three-phase total power.

(2) Measure the power of Y-connected load by two-wattmeter method. The connection of the power meter refers to Figure 16 - 2. Adjust the output line voltage of the voltage regulator from 0 to 220V slowly, measure the data according to Table 16 - 4 and calculate the three-phase total power.

Table 16 - 3　Power of Y_0-connected Three - Phase Load

Load Condition	Number Of Lights Per Phase	Measured Data			Calculated Value
		P_U (W)	P_V (W)	P_W (W)	ΣP (W)
Y_0-connected Symmetrical Load	3、3、3				
Y_0-connected Asymmetrical Load	1、2、3				

Table 16 - 4 Power of Y-connected Three-Phase Load

Load Condition	Number Of Lights Per Phase	Measured Data		Calculated Value
		P_1 (W)	P_2 (W)	ΣP (W)
Y-connected Symmetrical Load	3、3、3			
Y-connected Asymmetrical Load	1、2、3			

3. Measurement of Voltages and Current of △-connected Three-Phase Load

Connect the incandescent lamps to △-connected shape according to Figure 16 - 4. Adjust the output three-phase line voltage of the three-phase voltage regulator to 150V. Measure the current of each phase and line through the current sampling sockets and the load voltage across each phase when the load is symmetrical and asymmetrical, and fill in Table 16 - 5. Compare the brightness of incandescent lamps in each phase.

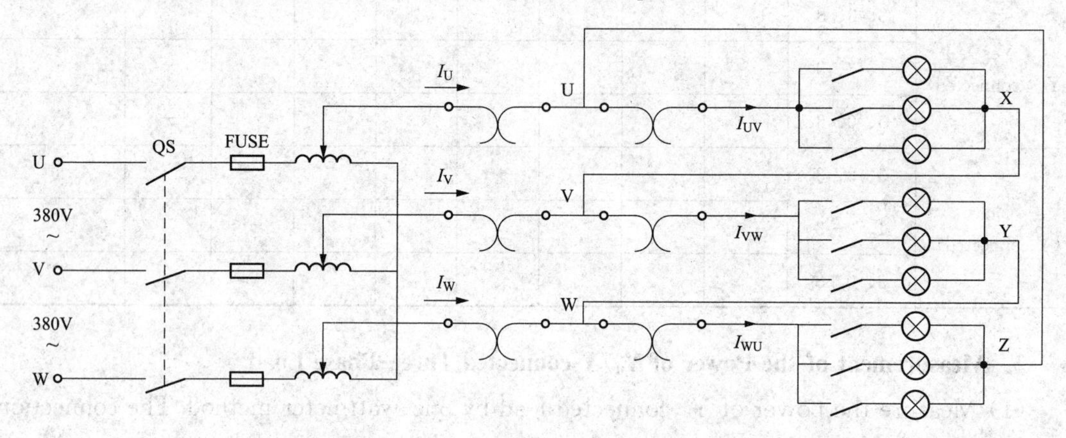

Figure 16 - 4 △-connect Three-Phase Load

Table 16 - 5 The Voltages and the Current of △-connected Three-Phase Load

Number of Lights Per Phase	Load Voltage (V)			Line Current (A)			Phase Current (A)			Brightness Comparison
	U_{UV}	U_{VW}	U_{WU}	I_U	I_W	I_V	I_{UV}	I_{VW}	I_{WU}	
3、3、3										
1、2、3										

4. Measurement of the Power of △-connected Three-Phase Load

Measure the power of △-connected three-phase load by two-wattmeter method. The connection of the power meter refers to Figure 16 - 2. Adjust the output line voltage of the voltage regulator from 0 to 150V slowly, measure the data according to Table 16 - 6 and calculate the three-phase total power.

● **Notes**

1. Before the start of the experiment, and every time the connection is changed, the handle of the three-phase regulator must be rotated to zero position.

Table 16 - 6 **Power of Δ-connected Three-Phase Load**

Load Condition	Number of Lights Per Phase	Measured Data		Calculated Value
		P_1 (W)	P_2 (W)	ΣP (W)
△-connected Symmetrical Load	3、3、3			
△-connected Asymmetrical Load	1、2、3			

2. The principles of experimental operation must be strictly observed: ① Connect the wire before the power is turned on when the experiment starts. ② Shut off the power supply before the wire is disconnected when the experiment ends.

3. Note the different sockets for measuring current when the line current and the phase current through △-connected load are measured.

4. Before the voltage is added to the circuit, turn on the switches of the lamps. The voltage should be monitored during the increasing of the voltage. If the Power supply voltage exceeds 100V but the lamps are still not lit, then adjust the voltage regulator to zero position and check the cause. Do not increase the voltage on the lamps too high to exceed the rated voltage.

● **Questions**

1. In this experiment, why is the line voltage of the three-phase power supply set to 220V in the Y-connected circuit?

2. Can a fuse be installed on the neutral wire? Why?

3. Why are there usually an ammeter and a voltmeter in the circuit when measuring power?

● **Experiment report**

1. According to the experiment data, on what conditions the formula $U_L = \sqrt{3}U_P$ is set up when the load is Y-connected? On what conditions the formula $I_L = \sqrt{3}I_P$ is set up when the load is △-connected?

2. Draw the phasor diagrams using the phase voltages of the symmetrical and asymmetrical Y-connected load. Summarize the role of the neutral wire of a three-phase four-wire power supply system using the experiment data and the observed phenomena.

3. Can an asymmetrical △-connected load works properly? Can the experiment prove this?

4. Draw the phasor diagram using the phase current of the asymmetrical △-connected load and calculate the line current according to the phasor diagram. Compare the calculated line current to the measured line current, analyze the result.

5. Compare the measurement results from one-wattmeter and two-wattmeter method. Summarize and analyze the methods and results of three-phase circuit power measurement.

实验 17　三相电路相序与无功功率的测量

一、实验目的

（1）掌握三相交流电路相序的测量方法。

（2）学会用功率表测量三相电路无功功率的方法。

二、实验原理

（1）相序指示器如图 17－1 所示，它是由一个电容器和两个白炽灯按 Y 形连接的电路，用来指示三相电源的相序。

在图 17－1 电路中，设 $\dot{U}_A$、$\dot{U}_B$、$\dot{U}_C$ 为三相对称电源相电压，中点电压为

$$\dot{U}_N=\frac{\dfrac{\dot{U}_A}{-jX_C}+\dfrac{\dot{U}_B}{R_B}+\dfrac{\dot{U}_C}{R_C}}{\dfrac{1}{-jX_C}+\dfrac{1}{R_B}+\dfrac{1}{R_C}}$$

设 $X_C=R_B=R_C$，$\dot{U}_A=U_P\angle 0°=U_P$ 代入上式得 $\dot{U}_N=(-0.2+j0.6)U_P$，

则
$$\dot{U}'_B=\dot{U}_B-\dot{U}_N=(-0.3-j1.466)U_P，U'_B=1.49U_P$$
$$\dot{U}'_C=\dot{U}_C-\dot{U}_N=(-0.3+j0.266)U_P，U'_C=0.4U_P$$

可见 $U'_B>U'_C$，B 相的白炽灯比 C 相的亮。

综上所述，用相序指示器指示三相电源相序的方法是如果连接电容器的一相是 A 相，那么，白炽灯较亮的一相是 B 相，较暗的一相是 C 相。

对于三相电动机而言，三相交流电接入电机后，相序决定了定子线圈形成的旋转磁场的方向，从而决定了转子的转动方向。当绕组相序 A、B、C 与电源相序 U、V、W 相同时，通电后，由电机轴伸端看，电机应顺时针方向旋转，反之则逆时针方向旋转。

（2）有功功率表不仅能测量有功功率，如果适当改换它的接线方式，也可用来测量无功功率。对于三相三线制供电的三相对称负载，可用一瓦特表法测得三相负载的总无功功 Q，测试原理线路如图 17－2 所示。

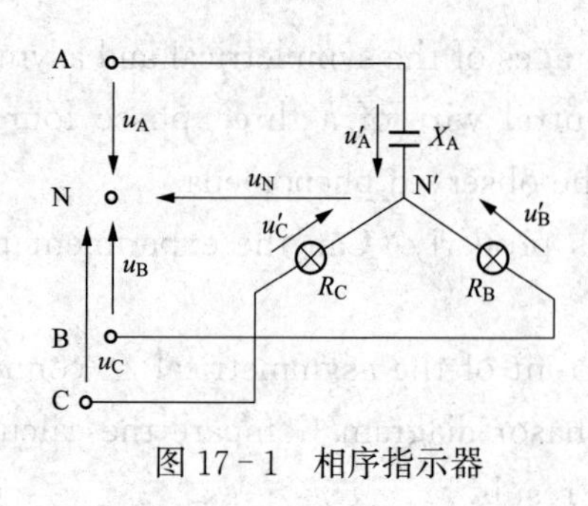

图 17－1　相序指示器

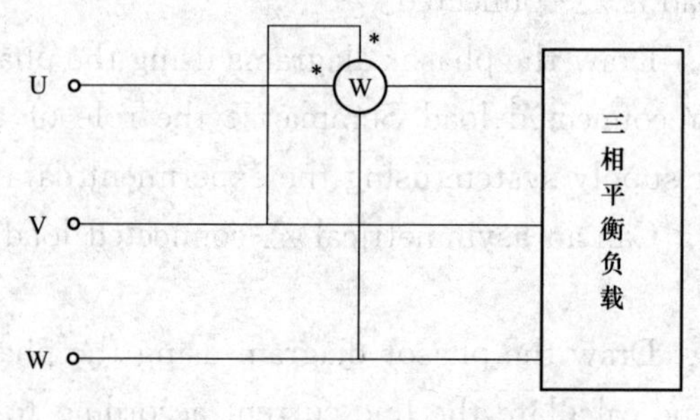

图 17－2　无功功率的测量

已知单相无功功率为

$$Q=UI\sin\varphi=UI\cos(90°-\varphi)$$

如果设法使加在电压线圈支路上的电压 $\dot{U}$ 与通过电流线圈的电流 $\dot{I}$ 之间的相位差等于

$(90°-\varphi)$，可将功率表用来测量无功功率。

将单相有功功率表的电流回路串入 U 相，电压回路（与电流同名端）端子接于 V 相，另一电压端子接于 W 相。则功率表的读数为

$$Q_1 = U_{VW} I_U \cos(90° - \varphi) = UI\sin\varphi$$

三相无功功率为

$$Q = \sqrt{3}Q_1 = \sqrt{3}UI\sin\varphi$$

故图 17-2 所示功率表读数的$\sqrt{3}$倍即为对称三相电路总的无功功率。功率表除了此图给出的一种连接法（I_U、U_{VW}）外，也可接成 I_V、U_{UW}或 I_W、U_{UV}。

三、实验设备

实验设备见表 17-1。

表 17-1　　实验设备

设备名称	型号与规格	数量	实验模块
智能仪表	0～500V 0～3A	3	NDG-01
电容	4.3μF/500V	1	NDG-08
白炽灯	220V	2	NDG-10
交流电源	0～450V 三相/0～250V 单相		QS-DYD3
三相笼型异步电动机	220V	1	M14

四、实验内容

1. 测定三相电源的相序

按图 17-1 接线，图中，$C=4.3\mu F$，R_B、R_C为两个 220V、25W 的白炽灯。调压器输出线电压为 220V 的三相交流电压。分别测量电容器、两个白炽灯到三相负载中点 N′的电压和三相电源中点 N 到负载中点 N′的电压（中点电压 u_N），观察灯光明暗状态并予以记录。设电容器一相为 A 相，试判断实际的 B、C 相。

2. 观察三相电机转动方向与相序的关系

异步电动机三相定子绕组的六个出线端有三个首端和三相末端。一般首端标以 A、B、C，末端标以 X、Y、Z。三相定子绕组可以接成 Y 形（连接 X-Y-Z）或△形（连接 A-Z、B-X、C-Y），如图 17-3(a)、(b) 所示，然后与三相交流电源相连。

由于电动机启动时 Y 形接法启动电流较小，故本步骤采用 Y 形接法。实验开始前，将调压器置于零位，如图 17-3(a) 所示将三相电源 U、V、W 与接成 Y 形的电动机连接，缓慢加压至三相电机刚好启动为止，观察并记录电机转动方向，与步骤 1 中测得相序互相对照验证。

3. 三相负载无功功率的测量

功率表接线如图 17-2 所示，三相负载为三相电动机。

(1) 将调压器置于零位，将三相电动机接成 Y 形接法。将线电压加至 220V，电动机运转稳定后，测量并记录无功功率。

(2) 将调压器回零，将三相电动机改接成△形接法，将线电压加至 150V，电动机运转稳定后，测量并记录无功功率。

根据所测无功功率数据，分别计算三相异步电动机总无功功率。

注意：切勿触碰电动机的转动部分。记录功率后即将电压回零。

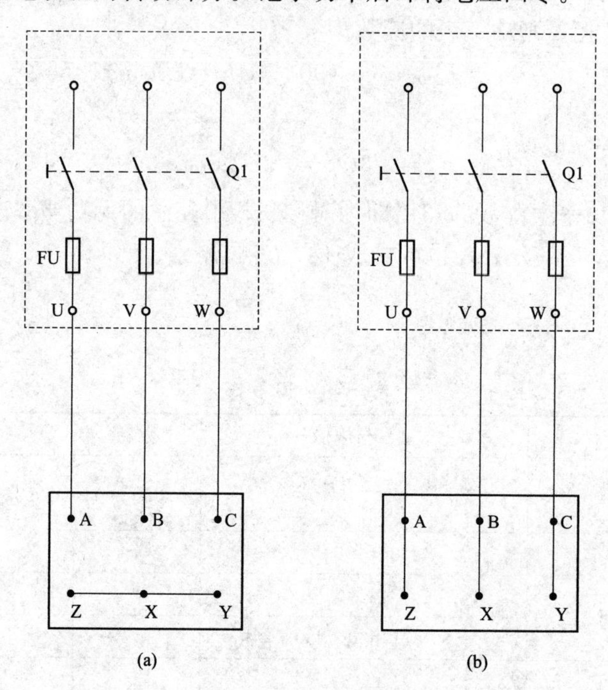

图 17－3　三相电动机的两种连接方式

（a）Y 形接法；（b）△形接法

五、注意事项

（1）实验之前，以及每次改变接线时，均必须将三相调压器调压手柄调至零位。

（2）必须严格遵守实验开始时先接线，后通电；实验结束时先断电，后拆线的实验操作原则。

六、思考题

在图 17－1 电路中，已知电源线电压为 220V，试计算电容器和白炽灯的电压，并与测量值相比较。

七、实验报告

对实验中观测到的现象进行归纳和总结。

Experiment 17 Measurement of Phase Sequence and Reactive Power of a Three-Phase Circuit

● **Objectives**

1. Learn how to measure phase sequence of a three-phase circuit.

2. Learn how to measure the reactive power of a three-phase circuit with a power meter.

● **Principles**

1. The phase-sequence indicator is shown in Figure 17 - 1. It's a circuit that is comprised of a capacitor and two incandescent lamps.

In the circuit of Figure 17 - 1, letting $\dot{U}_A$, $\dot{U}_B$, $\dot{U}_C$ as the phase voltages of the three-phase symmetrical power supply, and the neutral point voltage is

$$\dot{U}_N = \frac{\dfrac{\dot{U}_A}{-jX_C} + \dfrac{\dot{U}_B}{R_B} + \dfrac{\dot{U}_C}{R_C}}{\dfrac{1}{-jX_C} + \dfrac{1}{R_B} + \dfrac{1}{R_C}}$$

Letting $X_C = R_B = R_C$, $\dot{U}_A = U_P \angle 0° = U_P$, substituting them into the equation above, there is

$$\dot{U}_N = (-0.2 + j0.6)U_P$$

Therefore $\dot{U}'_B = \dot{U}_B - \dot{U}_N = (-0.3 - j1.466)U_P$, $U'_B = 1.49U_P$

$\dot{U}'_C = \dot{U}_C - \dot{U}_N = (-0.3 + j0.266)U_P$, $U'_C = 0.4U_P$

We get $U'_B > U'_C$, the incandescent lamp in B-phase is brighter than the one in C-phase.

In summary, the method of using phase sequence indicator to indicate phase sequence of a three-phase power supply is: if the phase connected to the capacitor is A-phase, then the phase with a brighter incandescent lamp is B-phase, and the phase with a darker lamp is C-phase.

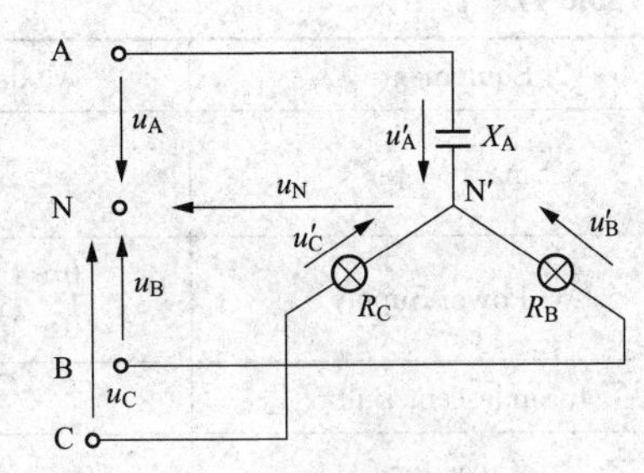

Figure 17 - 1 Phase-Sequence Indicator

For a three-phase motor, when it is connected to a three-phase AC power supply, the direction of the rotating magnetic field formed in the stator coil is determined by the phase sequence, thus the rotation direction of the rotor is determined. When the phase sequence of the winding (A, B, C) is the same as the phase sequence of the power supply (U, V, W), the motor will rotate clockwise while observed from shaft end , otherwise the motor rotates counterclockwise.

2. An active power meter not only can measure the active power, but also can be used to measure the reactive power if the connection is changed properly. To a symmetrical three-phase load supplied by a three-phase three-wire power supply, the one-wattmeter method

can be used to measure the total reactive power of three-phase load, the principle of connection is shown in Figure 17 - 2.

The single phase reactive power is

$$Q = UI\sin\varphi = UI\cos(90° - \varphi)$$

If the phase difference between the voltage $\dot{U}$ across the branch of the voltage coil and the current $\dot{I}$ through the current coil is $90° - \varphi$, the power meter can be used to measure reactive power.

Connect the current loop of the single-phase active power meter into phase U in series, and connect the dotted terminal of the voltage loop corresponding to the current loop to V-phase. Connect another voltage terminal to W-phase. The reading of the power meter is

$$Q_1 = U_{VW} I_U \cos(90° - \varphi) = UI\sin\varphi$$

The three phase active power is

$$Q = \sqrt{3}Q_1 = \sqrt{3}UI\sin\varphi$$

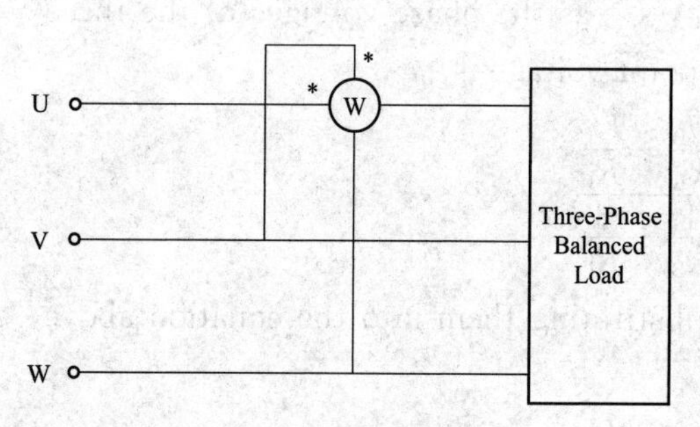

Figure 17 - 2 Measure of Reactive Power

Multiply the readings of the power meter in the Figure 17 - 2 by $\sqrt{3}$, we get the total reactive power of the symmetrical three-phase circuit. Besides the connection shown in this figure (I_U, U_{VW}), the power meter also can be connected as I_V, U_{UW} or I_W, U_{UV}.

● Equipment

Equipment is shown in Table 17 - 1.

Table 17 - 1 Equipment

Equipment	Model or Specification	Quantity	Module
Smart Meter	0～500V 0～3A	3	NDG - 01
AC Power Supply	0～450V Three-Phase 0～250V Single-Phase		QS - DYD3
Incandescent Bulb	25W	9	NDG - 10
Capacitor	4. 3μF/500V	1	NDG - 08
Three-Phase Squirrel-Cage Asynchronous Motor	220V	1	M14

● Contents

1. Determination of Phase Sequence of a Three-phase Power Supply

The experiment circuit is shown in Figure 17 - 1, where C=4. 3μF, R_B and R_C are two incandescent lamps of 220V/25W. The output line voltage of the voltage regulator is 220V. Measure the voltage across the capacitor and the neutral point of the three-phase load, the voltages across two incandescent lamps and N', and the neutral point voltage $U_{NN'}$. Observe the brightness of the two lamps and write them down. Letting the phase with the capacitor is

A-phase, try to determine actual B-phase and C-phase.

2. Observation of the Relationship between the Rotation Direction of the Three-Phase Motor and Phase Sequence of Power Supply

The six outlet terminals of three-phase stator winding of an asynchronous motor have three heads and three ends. Usually the heads are marked A, B, C, and the ends are marked X, Y, Z. The three-phase stator winding can be Y-connected or △-connected, as shown in Figure 17 - 3(a) and (b), then connected to the three-phase AC power supply.

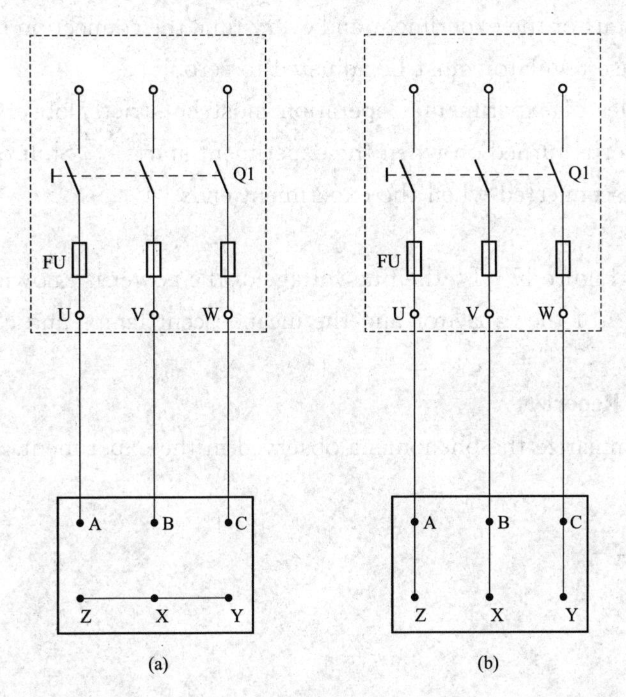

Figure 17 - 3 Two Ways of Connection for Three-Phase Motor
(a) Y-connected; (b) △-Connected

Because the starting current of the Y-connected motor is smaller, the Y-connected connection is used in this step. Before the experiment starts, adjust the output of the voltage regulator to zero. Connect the terminal U, V, W of the three-phase AC power supply to the motor according to Figure 17 - 3(a), and increase the voltage slowly until the motor is just starting to rotate. Observe and write down the rotating direction of the motor, compare and verify the results with the phase sequence measured in step 1.

3. Measurement of Reactive Power of Three-Phase Load

The connection of the power meter is shown in Figure 17 - 2. The load is the three-phase motor.

(1) Adjust the voltage regulator to zero, and connect the three-phase motor as Y-connected. Adjust the line voltage to 220V, measure and write down the reactive power when the rotating of the motor is stable.

(2) Adjust the voltage regulator to zero again, and change the connection of the three-

phase motor to △-connected. Adjust thc linc voltage to 150V, measure and write down the reactive power when the rotating of the motor is stable.

Calculate the total reactive power of three-phase asynchronous motor using the measured reactive power data separately.

In this experiment, do not touch the rotating rotor. Adjust the voltage to zero after writing down the reactive power.

● **Notes**

1. Before the start of the experiment and every time the connection is changed, the output of the three phase regulator must be adjusted to zero.

2. The principles of experimental operation must be strictly observed: ①Connect the wire before the power is turned on when the experiment starts. ②Shut off the power supply before the wire is disconnected when the experiment ends.

● **Questions**

In the circuit of Figure 17 - 1, the line voltage of the power is known to be 220V, try to calculate the voltages of the capacitor and the incandescent lamps and compare them to the measured values.

● **Experiment Report**

Sum up and summarize the phenomena observed in the experiment.

实验18　二端口网络的研究

一、实验目的

（1）加深理解二端口网络的基本理论。

（2）掌握直流二端口网络传输参数的测试方法。

二、实验原理

1. 二端口网络的基本概念

对于任何一个线性二端口网络，通常被关心的往往只是输入端口和输出端口电压和电流间的相互关系。二端口网络端口的电压和电流四个变量之间的关系，可以用多种形式的参数方程来表示。本实验采用输出口的电压U_2和电流I_2作为自变量，以输入口的电压U_1和电流I_1作为应变量，所得的方程称为二端口网络的传输方程。如图18-1所示的无源线性二端口网络（又称为四端网络）的传输方程为

$$U_{1S}=AU_{2S}+B(-I_{2S})$$
$$I_{1S}=CU_{2S}+D(-I_{2S})$$

式中：A、B、C、D为二端口网络的传输参数，其值完全决定于网络的拓扑结构及各支路元件的参数值。这四个参数表征了该二端口网络的基本特性。

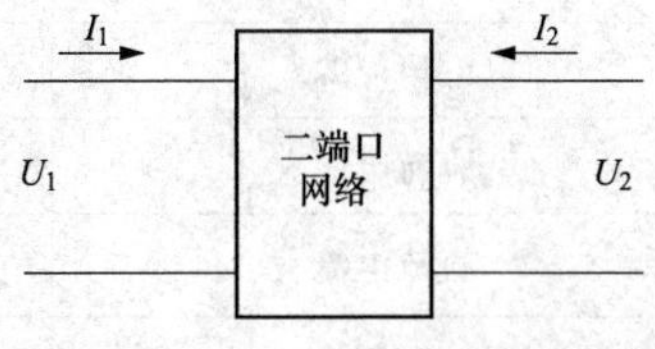

图18-1　二端口网络

2. 二端口网络传输参数的测试方法

（1）双端口同时测量法。在网络的输入口加上电压，在两个端口同时测量其电压和电流，由传输方程可得A、B、C、D四个参数，有

$A=\dfrac{U_{1O}}{U_{2O}}$（令$I_2=0$，即输出口开路时）　$B=\dfrac{U_{1S}}{-I_{2S}}$（令$U_2=0$，即输出口短路时）

$C=\dfrac{I_{1O}}{U_{2O}}$（令$I_2=0$，即输出口开路时）　$D=\dfrac{I_{1S}}{-I_{2S}}$（令$U_2=0$，即输出口短路时）

（2）双端口分别测量法。在输入口加电压，将输出口开路和短路，测量输入口的电压和电流，由传输方程可得

$$R_{1O}=\frac{U_{1O}}{I_{1O}}=\frac{A}{C}\text{（令}I_2=0\text{，即输出口开路时）}$$

$$R_{1S}=\frac{U_{1S}}{I_{1S}}=\frac{B}{D}\text{（令}U_2=0\text{，即输出口短路时）}$$

然后在输出口加电压，将输入口开路和短路，测量输出口的电压和电流，由传输方程可得$R_{2O}=\dfrac{U_{2O}}{I_{2O}}=\dfrac{D}{C}$（令$I_1=0$，即输入口开路时）

$$R_{2S}=\frac{U_{2S}}{I_{2S}}=\frac{B}{A}\text{（令}U_1=0\text{，即输入口短路时）}$$

R_{1O}、R_{1S}、R_{2O}、R_{2S}分别表示一个端口开路和短路时另一端口的等效输入电阻，这四个参数中有三个是独立的，因此，只要测量出其中任意三个参数（如R_{1O}、R_{2O}、R_{2S}），与方

程 $AD-BC=1$（二端口网络为互易双口，该方程成立）联立，便可求出四个传输参数

$$A=\sqrt{R_{1O}/(R_{2O}-R_{2S})},\ B=R_{2S}A,\ C=A/R_{1O},\ D=R_{2O}C$$

3. 二端口网络的级联

二端口网络级联后的等效二端口网络的传输参数也可采用上述方法之一求得。根据二端口网络理论推得：二端口网络 1 与二端口网络 2 级联后等效的二端口网络的传输参数，与网络 1 和网络 2 的传输参数之间有如下的关系

$$A=A_1A_2+B_1C_2$$
$$B=A_1B_2+B_1D_2$$
$$C=C_1A_2+D_1C_2$$
$$D=C_1B_2+D_1D_2$$

三、实验设备

实验设备见表 18－1。

表 18－1　实　验　设　备

设备名称	型号与规格	数量	实验模块
恒压源	0～30V	1	NDG－02
直流电压表	0～200V	1	NDG－03
直流电流表	0～2000mA	1	
万用表	UT803	1	NDG－13
电阻	200Ω	1	
电阻	300Ω	1	
电阻	510Ω	1	
可调电阻	1～9999Ω	2	NDG－06
电位器	1kΩ	1	NDG－12

四、实验内容

实验电路如图 18－2 所示，其中图 18－2(a) 为 T 型网络，图 18－2(b) 为 Π 型网络。将恒压源的输出电压调到 10V，作为二端口网络的输入电压。各个电流均用电流插头、取样插座测量。

1. 用双端口同时测量法测定二端口网络传输参数

根据双端口同时测量法的原理和方法，按照表 18－2、表 18－3 的内容，分别测量 T 型网络和 Π 型网络的电压、电流，并计算出传输参数 A、B、C、D 值，将所有数据记入表中。

2. 用双端口分别测量法测定级联二端口网络传输参数

将 T 型网络的输出口与 Π 型网络的输入口连接，组成级联二端口网络（数量不足的电阻使用可调电阻及电位器代替）。根据双端口分别测量法的原理和方法，按照表 18－4 的内容，分别测量级联二端口网络输入口和输出口的电压、电流，并计算出等效输入电阻和传输参数 A、B、C、D，将所有数据记入表中。

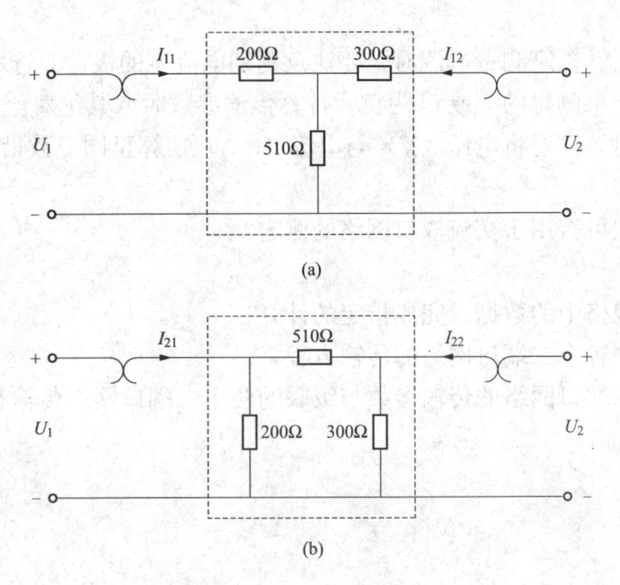

图 18－2　二端口网络实验电路

(a) T 型网络；(b) Ⅱ 型网络

表 18－2　　测定 T 型网络传输参数

		测量值			计算值	
T 型网络	输出端开路 ($I_2=0$)	U_{1O} (V)	U_{2O} (V)	I_{1O} (mA)	A	C
	输出端短路 ($U_2=0$)	U_{1S} (V)	I_{1S} (mA)	I_{2S} (mA)	B	D

表 18－3　　测定 Ⅱ 型网络传输参数

		测量值			计算值	
Ⅱ 型网络	输出端开路 ($I_2=0$)	U_{1O} (V)	U_{2O} (V)	I_{1O} (mA)	A	C
	输出端短路 ($U_2=0$)	U_{1S} (V)	I_{1S} (mA)	I_{2S} (mA)	B	D

表 18－4　　T 测定级联二端口网络传输参数

输出端开路 ($I_2=0$)			输出端短路 ($U_2=0$)			计算值	
U_{1O} (V)	I_{1O} (mA)	R_{1O}	U_{1S} (V)	I_{1S} (mA)	R_{1S}	A	C
输入端开路 ($I_1=0$)			输入端短路 ($U_1=0$)				
U_{2O} (V)	I_{2O} (mA)	R_{2O}	U_{2S} (V)	I_{2S} (mA)	R_{2S}	B	D

五、注意事项

(1) 用电流插头、插座测量电流时，要注意判别电流表的极性及选取适合的量程。

(2) 实验中，如果测得的 I 或 U 为负，计算传输参数时取其绝对值。

(3) 计算参数时，注意将电压（V）与电流（mA）换算至同一数量级进行计算。

六、思考题

本实验中的方法可否用于交流双口网络的测定？

七、实验报告

(1) 整理各个表格中的数据，完成指定的计算。

(2) 写出实验中各个二端口网络的传输方程。

(3) 验证级联二端口网络的传输参数与级联的两个二端口网络传输参数之间的关系。

Experiment 18 Study of Two-Port Circuits

● **Objectives**

1. Acquire a better understanding of the basic theory of two-port circuits.
2. Learn how to test the transmission parameters of DC two-port circuits.

● **Principles**

1. Basic Concepts of Two-Port Circuits

For any linear two-port circuit, it is usually just the interrelationship between the voltages and the current of the input port and the output port is concerned. The relationships between the four variables of voltage and current of the ports of the two-port circuit can be expressed as parametric equations in a variety of forms. In this experiment, the voltage U_2 and the current I_2 of the output port are used as independent variables, and the voltage U_1 and the current I_1 of the input port are used as dependent variables, the derived equation is called the transmission equation of the two-port circuit. The transmission equations of the passive linear two port circuit (aka four-terminal circuit) shown in Figure 18 - 1 are

$$U_{1S} = AU_{2S} + B(-I_{2S})$$
$$I_{1S} = CU_{2S} + D(-I_{2S})$$

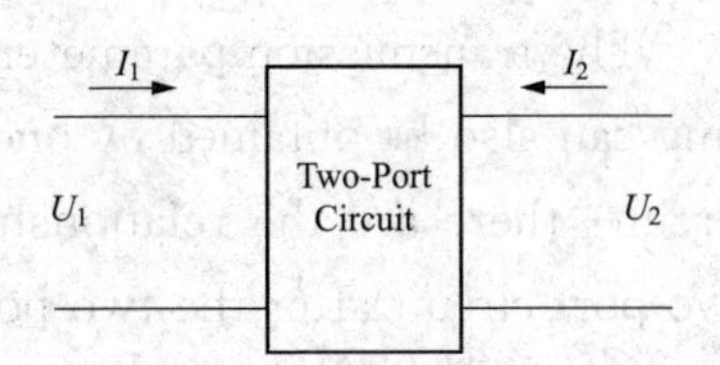

Figure 18 - 1 Two-Port Circuit

A, B, C, D in the equations are the transmission parameters of the two-port circuit, the values of which are completely determined by the topology of the circuit and the parameters of elements in each branch. These four parameters represent the basic characteristics of the two-port circuit.

2. Test Methods for Transmission Parameters of Two-Port Circuit

(1) Simultaneous Measurement in Two Ports of the Circuit

Add a voltage to the input port of the circuit, and measure the voltages and the current simultaneously at the two ports. The four parameters A, B, C and D can be obtained from the transmission equations:

$$A = \frac{U_{1O}}{U_{2O}}$$ (letting $I_2=0$, that is, when the output port is open-circuited)

$$B = \frac{U_{1S}}{-I_{2S}}$$ (letting $U_2=0$, that is, when the output port is short-circuited)

$$C = \frac{I_{1O}}{U_{2O}}$$ (let $I_2=0$, that is, when the output port is open-circuited)

$$D = \frac{I_{1S}}{-I_{2S}}$$ (letting $U_2=0$, that is, when the output port is short-circuited)

(2) Separate Measurement in Two Ports of the Circuit

Add a voltage to the input port of the circuit, open and short the output port, and measure the voltage and the current of the input port. From the transmission equations, there are

$$R_{1O}=\frac{U_{1O}}{I_{1O}}=\frac{A}{C}\text{ (letting } I_2=0\text{, that is, when the output port is open-circuited)}$$

$$R_{1S}=\frac{U_{1S}}{I_{1S}}=\frac{B}{D}\text{ (letting } U_2=0\text{, that is, when the output port is short-circuited)}$$

Then add a voltage to the output port, open and short the input port, and measure the voltage and the current of the output port. From the transmission equations, there are

$$R_{2O}=\frac{U_{2O}}{I_{2O}}=\frac{D}{C}\text{ (letting } I_1=0\text{, that is, when the input port is open-circuited)}$$

$$R_{2S}=\frac{U_{2S}}{I_{2S}}=\frac{B}{A}\text{ (letting } U_1=0\text{, that is, when the input port is short-circuited)}$$

R_{1O}, R_{1S}, R_{2O}, R_{2S} are the equivalent input resistances of one port when another port is open-circuited and short-circuited. Three of these four parameters are independent, therefore, as long as any three of the parameters are measured, and coupled with the equation $AD-BC=1$ (the two-port circuit is reciprocal, the equation is established), the transmission parameters can be obtained

$$A=\sqrt{R_{1O}/(R_{2O}-R_{2S})},\quad B=R_{2S}A,\quad C=A/R_{1O},\quad D=R_{2O}C$$

3. The Cascade of Two-Port Circuits

The transmission parameters of equivalent two-port circuit after two-port circuits cascading can also be obtained by one of the above methods. Derived from the theory of two-port circuit: there are the relationships between the transmission parameters of the equivalent two-port circuit after the two-port circuit 1 and the two-port circuit 2 cascade and the transmission parameters of circuit 1 and circuit 2

$$A=A_1A_2+B_1C_2$$
$$B=A_1B_2+B_1D_2$$
$$C=C_1A_2+D_1C_2$$
$$D=C_1B_2+D_1D_2$$

● Equipment

Equipment is shown in Table 18 - 1.

Table 18 - 1 Equipment

Equipment	Model or Specification	Quantity	Module
Constant Voltage Source	0～30V	1	NDG - 02
DC Voltmeter	0～200V	1	NDG - 03
DC Ammeter	0～2000mA	1	
Multimeter	UT803	1	
Resistor	200Ω	1	NDG - 13
	300Ω	1	
	510Ω	1	

续表

Equipment	Model or Specification	Quantity	Module
Adjustable Resistance	1～9999Ω	2	NDG－06
Potentiometer	1 kΩ	1	NDG－12

● Contents

The experiment circuits are shown in Figure 18－2. Figure 18－2(a) is the T-shape circuit, and Figure 18－2(b) is the Π-shape circuit. Adjust the output of the constant voltage source to 10V and use it as the input of the two-port circuit. Measure the current with current plugs and current sampling sockets.

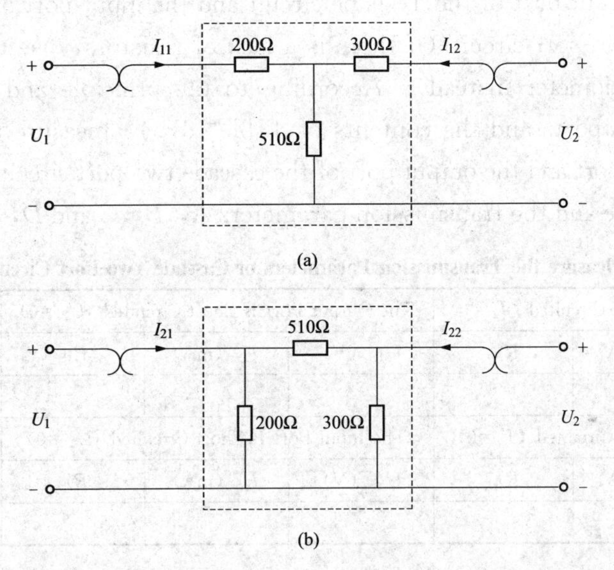

Figure 18－2 The Experiment Circuit for Two-Port Circuit Experiment
(a)T-Shape Circuit; (b)Π-Shape Circuit

1. Measure the Transmission Parameters of the Two-Port Circuit by Simultaneous Measurement in Two Ports

According to the principle and method of simultaneous measurement in two ports of the circuit and the contents of Table 18－2 and Table 18－3, measure the voltages and the current of the T-shape circuit and the Π-shape circuit. Calculate the transmission parameters A, B, C and D, and fill in the tables.

Table 18－2 Measure the Transmission Parameters of the T-Shape Two-Port Circuit

<table>
<tr><td rowspan="4">T-Shape Circuit</td><td rowspan="2">The Output Port is Open-Circuited ($I_2=0$)</td><td colspan="3">Measured Value</td><td colspan="2">Calculated Value</td></tr>
<tr><td>U_{10} (V)</td><td>U_{20} (V)</td><td>I_{10} (mA)</td><td>A</td><td>C</td></tr>
<tr><td rowspan="2">The Output Port Is Short-Circuited ($U_2=0$)</td><td>U_{1S} (V)</td><td>I_{1S} (mA)</td><td>I_{2S} (mA)</td><td>B</td><td>D</td></tr>
<tr><td></td><td></td><td></td><td></td><td></td></tr>
</table>

Table 18 - 3　　Measure the Transmission Parameters of the Π-Shape Two-Port Circuit

Π-Shape Circuit	The Output Port is Open-Circuited ($I_2=0$)	Measured Value			Calculated Value	
		U_{1O} (V)	U_{2O} (V)	I_{1O} (mA)	A	C
	The Output Port Is Short-Circuited ($U_2=0$)	U_{1S} (V)	I_{1S} (mA)	I_{2S} (mA)	B	D

2. Measure the Transmission Parameters of Cascade Two-Port Circuit by Separate Measurement in Two Ports of the Circuit

Connect the output port of the T-shape circuit and the input port of the Π-shape circuit to form a cascade two-port circuit (if there is a lack of resistors, use the adjustable resistance and the potentiometer instead). According to the principle and method of separate measurement in two ports and the contents of Table 18 - 4, measure the voltages and the current of the input port and the output port of the cascade two-port circuit. Calculate the equivalent input resistance and the transmission parameters A, B, C and D, and fill in the table.

Table 18 - 4　　Measure the Transmission Parameters of Cascade Two-Port Circuit

The Output Port Is Open-Circuited ($I_2=0$)			The Output Port Is Short-Circuited ($U_2=0$)			Calculated Value	
U_{1O} (V)	I_{1O} (mA)	R_{1O}	U_{1S} (V)	I_{1S} (mA)	R_{1S}	A	C
The Input Port Is Open-Circuited ($I_1=0$)			The Input Port Is Short-Circuited ($U_1=0$)				
U_{2O} (V)	I_{2O} (mA)	R_{2O}	U_{2S} (V)	I_{2S} (mA)	R_{2S}	B	D

- **Notes**

1. Pay attention to the polarity and the range of the ammeter while measuring the current with the current plugs and sockets.

2. In the process of the experiment, if a measured current or voltage is negative, use the absolute value of which for the calculation of the transmission parameters.

3. While calculating the parameters, convert the voltages (V) and the current (mA) to the same order of magnitude.

- **Questions**

Can the methods in this experiment be used for the measurement of an AC two-port circuit?

- **Experiment Report**

1. Sort out the data in the tables, finish the designated calculations.

2. Write the transmission equations of the two-port circuits in the experiment.

3. Verify the relationships between the transmission parameters of cascade two-port circuit and the transmission parameters of the two cascaded two-port circuits.

实验 19　裂相电路的研究

一、实验目的

（1）了解三相异步电动机的启动与运行。

（2）掌握单相交流电源转换为三相交流电源的原理。

（3）掌握单相电源驱动三相异步电动机的方法。

二、实验原理

1. 裂相基本原理

单相裂相是将单相交流电源利用电阻、电感、电容的移相原理得到三相电源。原理如图 19－1 所示，当 $X_L = X_C = \sqrt{3}R$ 时，Y 形接法的电阻 R 上的电压为对称三相电压。相量图如图 19－2 所示。

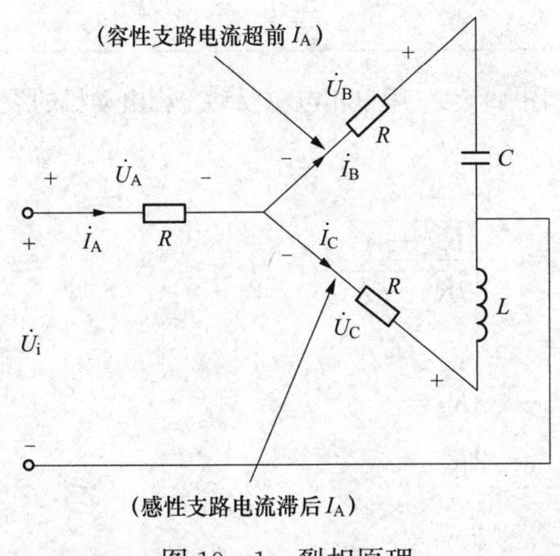

图 19－1　裂相原理

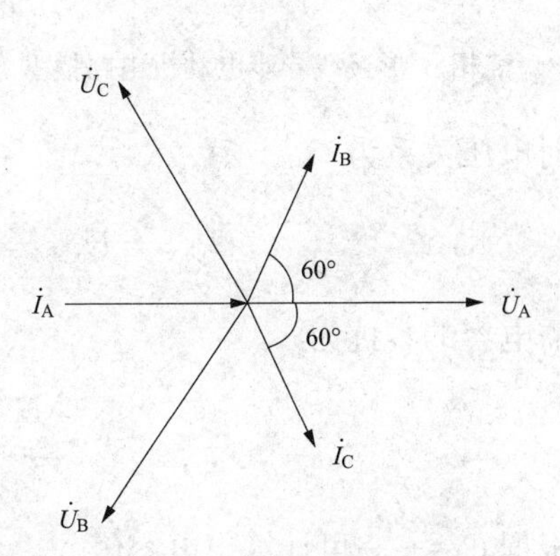

图 19－2　裂相电路相量图

当 $X_L = X_C = \sqrt{3}R$，根据原理图可写出下列各式

$$\dot{I}_B = \frac{R + jX_L}{(R + jX_L) + (R - jX_C)}\dot{I}_A = \left(\frac{1}{2} + j\frac{\sqrt{3}}{2}\right)\dot{I}_A$$

$$\dot{I}_C = \frac{R - jX_C}{(R + jX_L) + (R - jX_C)}\dot{I}_A = \left(\frac{1}{2} - j\frac{\sqrt{3}}{2}\right)\dot{I}_A$$

$$\dot{U}_A = R\dot{I}_A = \dot{U}_A \angle 0^\circ$$

$$\dot{U}_B = -R\dot{I}_B = \dot{U}_A \angle 120^\circ$$

$$\dot{U}_C = -R\dot{I}_C = \dot{U}_A \angle 120^\circ$$

$$\dot{U}_A = \dot{U}_B = \dot{U}_C = \frac{1}{3}U_i$$

2. 三相电机负载裂相原理

三相 Y 形接线异步电动机的电路模型如图 19－3 所示。三相异步电动机需要对称三相电源才能正常运转。若要用单相电源驱动三相电动机，考虑到对于实际电动机而言，一般有

$X_L > \sqrt{3}R$，无法满足 $X_L = \sqrt{3}R$ 的条件，因此可在其中两相（如 B、C）串接电容，如图 19-4 所示。

在图 19-4 中，支路 B 接入的电容应满足 $X_{C2} - X_L = \sqrt{3}R$，支路 C 接入的电容应满足 $X_L - X_{C3} = \sqrt{3}R$。此时支路 B 呈容性，支路 C 呈感性，且满足 $X_{eq} = \sqrt{3}R$ 的条件，从而实现单相转为对称三相的目的。

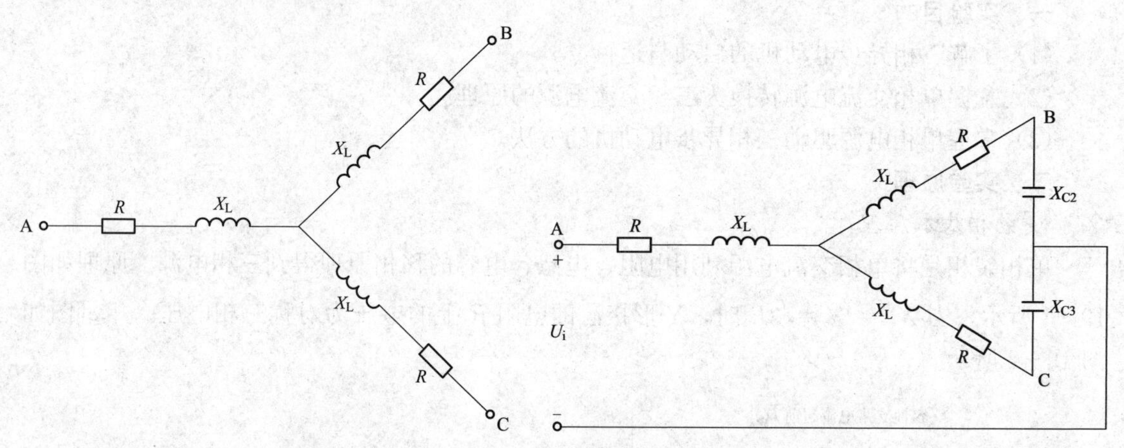

图 19-3 三相 Y 形接线异步电动机电路模型

图 19-4 用单相电源驱动三相电动机的裂相电路

此时电压关系为

$$U_{phA} = U_{phB} = U_{phC} = \frac{\sqrt{R^2 + X_L^2}}{\sqrt{9R^2 + X_L^2}} U_i$$

所需电容的容抗为

$$X_{C2} = X_L + \sqrt{3}R$$
$$X_{C3} = X_L - \sqrt{3}R$$

再根据 $C = \dfrac{1}{\omega X_C}$ 可计算出电容值。

$X_L < \sqrt{3}R$ 时，支路中电容 C_3 应改为电感 L_3，相关算式请自行推导。

三、实验设备

实验设备见表 19-1。

表 19-1 实 验 设 备

设备名称	型号与规格	数量	实验模块
智能仪表	0～500V 0～3A	3	NDG-01
交流电源	0～450V 三相/0～250V 单相		QS-DYD3
三相笼型异步电动机	220V	1	M14
可调电容电感箱		1	
钳子		1	

四、实验内容

(1) 观察不同相数电源供电下电动机的启动状况。将调压器按以下要求接入 Y 形三相异步电动机：①单相电源为电动机供电；②二相电源为电动机供电；③三相对称电源为电动机供电。用交流电压表监测调压器输出，每次均从 0 开始缓慢加压，至电动机启动或电源线电压达 380V 仍不能启动为止。分别观察记录 3 种情况下电动机能否启动。

(2) 测量三相异步电动机在对称三相电源供电下的启动（堵转）绕组参数。在对称三相电源供电的情况下，将交流电压表、交流电流表、功率表按二表法接入电路，接线如图 19－5 所示。用钳子夹住电机转轴（堵转），将三相调压器输出从 0 缓慢加压，直至电流达 $I=0.6I_N$（I_N为电动机的额定电流，可在电动机铭牌上查找），测量线电压和功率，将数据记入表 19－2。

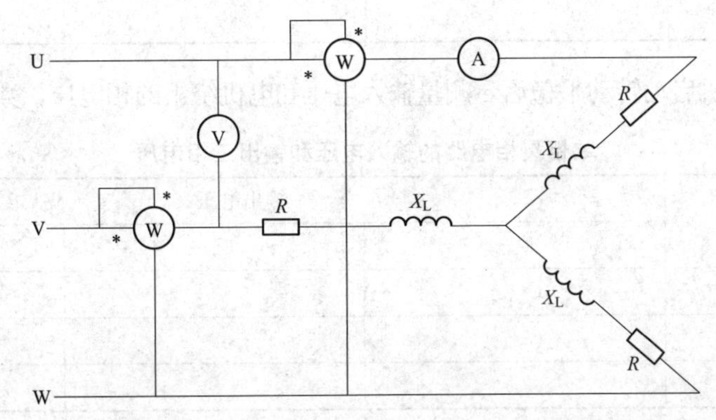

图 19－5 测量三相异步电动机在对称三相电源供电下的启动参数

表 19－2 电动机在对称三相电源供电下启动（堵转）时的单相参数

<table>
<tr><td>测量</td><td>$I=0.6I_N$（A）</td><td>U_{UV}（V）</td><td>P_{UW}（W）</td><td>P_{VW}（W）</td></tr>
<tr><td></td><td></td><td></td><td></td><td></td></tr>
<tr><td rowspan="4">计算</td><td colspan="2">$\sum P$（W）</td><td colspan="2"></td></tr>
<tr><td colspan="2">每相功率 P（W）</td><td colspan="2"></td></tr>
<tr><td colspan="2">每相电阻 R（Ω）</td><td colspan="2"></td></tr>
<tr><td colspan="2">每相感抗 X（Ω）</td><td colspan="2"></td></tr>
</table>

其中：电阻 $R=P/I^2$，$X=\sqrt{(U/I)^2-R^2}$（U 为相电压）。

(3) 计算串入电动机绕组的电容（电感）值。根据表 19－2 的 R 和 X 参数计算所串入电容（电感）的容抗（感抗）。

若 $X_L>\sqrt{3}R$，则

$$X_{C2}=X_L+\sqrt{3}R$$
$$X_{C3}=X_L-\sqrt{3}R$$

若 $X_L<\sqrt{3}R$，则其中一相应串入补偿电感，有

$$X_{C2}=X_L+\sqrt{3}R$$
$$X_{L3}=\sqrt{3}R-X_L$$

再根据 $C=\dfrac{1}{\omega X_C}$，$L_3=\dfrac{X_{L3}}{\omega}$进行计算，记录计算结果。

(4) 将 C_2接入电动机 B 相，C_3（或 L_3）接入电动机 C 相，加单相电源，缓慢调节调压器，观察电动机是否转动，转动是否稳定。若稳定程度不好，说明三相不够对称，需要调整

电容值。重新测定电动机参数，重新选择裂相所需的元件、参数，直到电动机旋转稳定，噪声最小。

（5）对裂相后的电路调压器回零，重新启动电动机，测量启动时电流、电压和功率，数据记入表 19－3，与三相电源启动时的值比较。

表 19－3　两种电源供电下电机的启动参数比较

	电流（A）	电压（V）	功率（W）
三相电源			
单相电源			
差　值			

（6）观察电动机启动转动平稳后，测量输入电压和电机绕组的相电压，数据记入表 19－4。

表 19－4　单相裂相电路的输入电压和输出三相电压

输入电压（V）	输出电压（V）		
	U_{AB}	U_{BC}	U_{CA}
40			
50			
60			

五、注意事项

（1）单相、二相、三相电源作用进行比较时，每次起始状态都是调压器回零，电动机静止，然后调压器缓慢升压，在开始转动（或电源线电压达 380V）时停止升压，记录状态。

（2）堵转状态需首先将调压器回零，在转轴静止时将其卡住，然后缓慢加压。禁止对卡住转子的电动机突然加入较高电压，严禁试图卡住转动的转子。

（3）本实验仅为验证电源可由单相转变为三相，不能用于电动机实际运行。

六、思考题

（1）三相电动机启动和运行时参数有何不同？

（2）在测电动机参数时，为什么不直接测相电压？为什么不使用三瓦计法测功率？

七、实验报告

（1）推导 $X_L < \sqrt{3}R$ 时裂相电路参数的计算公式。

（2）对裂相实验的结果进行分析，总结裂相电路的特点。

Experiment 19 Study of Splitting Phase Circuit

- **Objectives**

1. Understand the starting and the running state of three-phase asynchronous motor.
2. Understand the principle of conversion of single-phase AC power supply to three-phase AC power supply.
3. Learn how to drive a three-phase asynchronous motor with a single-phase power supply.

- **Principles**

1. The Basic Principle of Splitting Phase

Single-phase splitting phase is using the phase shifting principle of the resistors, the inductors and the capacitors to obtain a three-phase power supply. The principle is shown in Figure 19－1, when $X_L = X_C = \sqrt{3}R$, the voltages on the Y-connected resistors are symmetrical three-phase voltages. The phasor graph is shown in Figure 19－2.

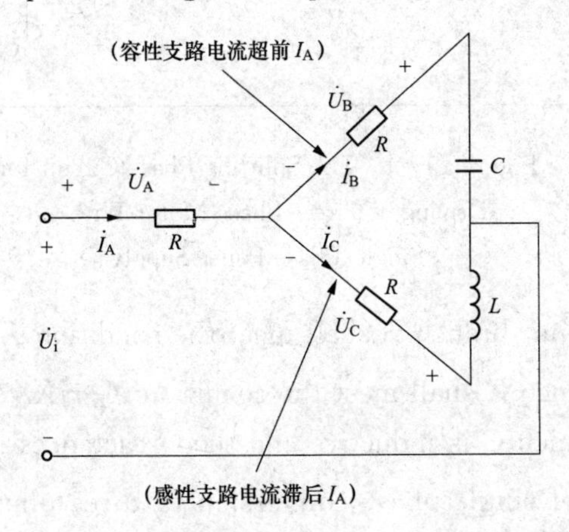

Figure 19－1 The Principle of Splitting Phase

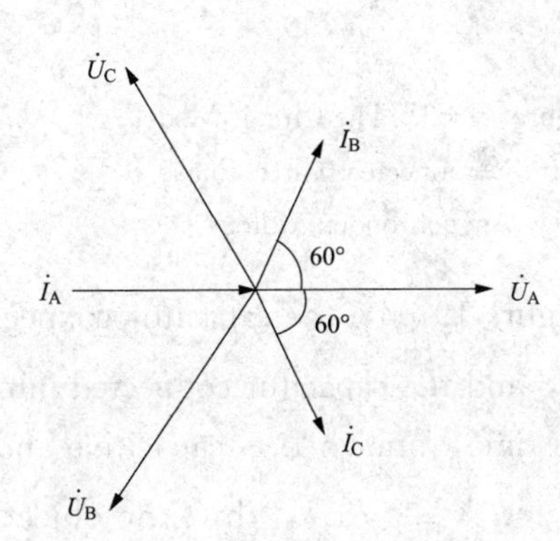

Figure 19－2 The Phase Graph of Splitting Phase Circuit

When $X_L = X_C = \sqrt{3}R$, the following equations can be written according to the principle figure

$$\dot{I}_B = \frac{R + jX_L}{(R + jX_L) + (R - jX_C)}\dot{I}_A = \left(\frac{1}{2} + j\frac{\sqrt{3}}{2}\right)\dot{I}_A$$

$$\dot{I}_C = \frac{R - jX_C}{(R + jX_L) + (R - jX_C)}\dot{I}_A = \left(\frac{1}{2} - j\frac{\sqrt{3}}{2}\right)\dot{I}_A$$

$$\dot{U}_A = R\dot{I}_A = \dot{U}_A \underline{/0^\circ}$$

$$\dot{U}_B = -R\dot{I}_B = \dot{U}_A \underline{/120^\circ}$$

$$\dot{U}_C = -R\dot{I}_C = \dot{U}_A \underline{/120^\circ}$$

$$\dot{U}_A=\dot{U}_B=\dot{U}_C=\frac{1}{3}U_i$$

2. The Splitting Phase Principle for a Three-Phase Motor Load

The circuit model of a Y-connected three-phase asynchronous motor is shown in Figure 19 - 3. A three-phase asynchronous motor needs a symmetrical three-phase power supply to operate properly. If a single-phase power supply is used to drive a three-phase motor, considering that for the actual motor, in general, there is $X_L > \sqrt{3}R$, and the condition of $X_L = \sqrt{3}R$ is unable to be met. Therefore, the series capacitors shall be connected to two phases (e. g. B and C), as shown in Figure 19 - 4.

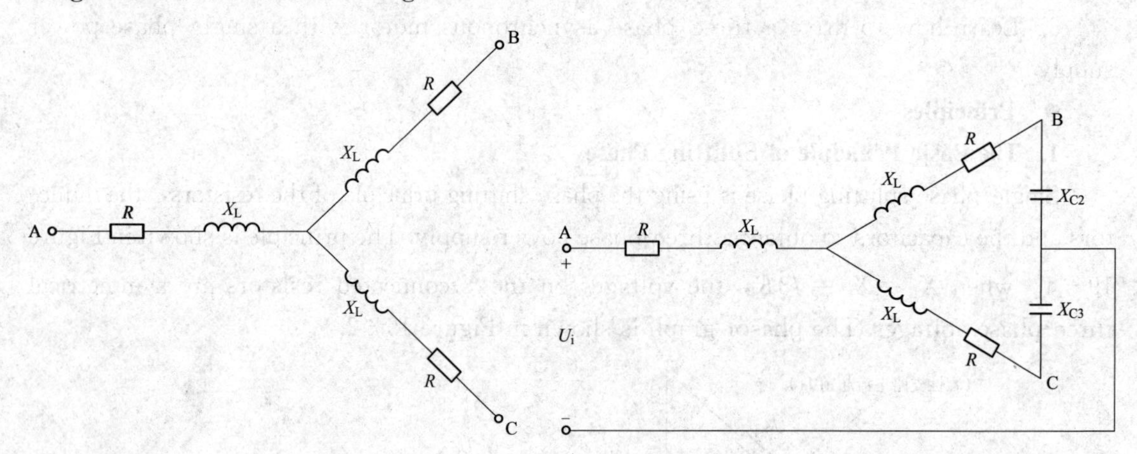

Figure 19 - 3 The Circuit Model of a Y-connected Three-Phase Asynchronous Motor

Figure 19 - 4 The Splitting Phase Circuit for Driving a Three -Phase Motor With a Single-Phase Pouer Supply

In Figure 19 - 4, the capacitor connected into branch B shall meet the condition $X_{C2} - X_L = \sqrt{3}R$, and the capacitor connected into branch C shall meet the condition $X_L - X_{C3} = \sqrt{3}R$. At this time, branch B is capacitive and branch C is inductive and their reactances meet the condition $X_{eq} = \sqrt{3}R$, thus the objective of single phase conversion to three phase is achieved.

At this time, the relationship between the voltages is

$$U_{phA}=U_{phB}=U_{phC}=\frac{\sqrt{R^2+X_L^2}}{\sqrt{9R^2+X_L^2}}U_i$$

The capacitive reactances of the required capacitors are

$$X_{C2}=X_L+\sqrt{3}R$$

$$X_{C3}=X_L-\sqrt{3}R$$

The capacitances can be calculated using $C=\frac{1}{\omega X_C}$.

When $X_L < \sqrt{3}R$, the capacitor C_3 in the branch shall be replaced by a inductor L_3. Please deduce the correlation equation.

● Equipment

Equipment is shown in Table 19 - 1.

Table 19 - 1　　Equipment

Equipment	Model or Specification	Quantity	Module
Smart Meter	0～500V 0～3A	3	NDG - 01
AC Power supply	0～450V Three-Phase 0～250V Single-Phase		QS - DYD3
Three-Phase Squirrel-Cage Asynchronous Motor	220V	1	M14
Adjustable Capacitance And Inductance Box		1	
Plier		1	

● Contents

1. Observe the Starting Conditions of the Motor under the Power Supply with Different Quantities of Phases

Connect the voltage regulator to the Y-connected three-phase to asynchronous motor according to the following requirements: ① single-phase power supply supplies to the motor; ② two-phase power supply supplies to the motor; ③ three-phase symmetrical power supply supplies to the motor. Using AC voltmeter to monitor the voltage regulator output, slowly increase the voltage from zero every time, until the motor starts or the motor is still unable to start when the line voltage of the power supply is up to 380V. Observe and write down whether the motor can start in these situations.

2. Measure the Starting (Locked Rotor) Winding Parameters of Three-Phase Asynchronous Motor under the Supply of Symmetrical Three-Phase Power Supply

Connect AC voltmeter, AC ammeter and power meter into the circuit according to the two-wattmeter method under the supply of symmetrical three-phase power supply, as shown in Figure 19 - 5. Clamp the motor rotor with a Pair of Plier (lock the rotor), and increase the output of the three-phase voltage regulator from zero slowly until the current reaches $I = 0.6I_N$. I_N is the rated current of the motor, it can be found on the nameplate of the motor. Measure the line voltage and power, and fill in Table 19 - 2.

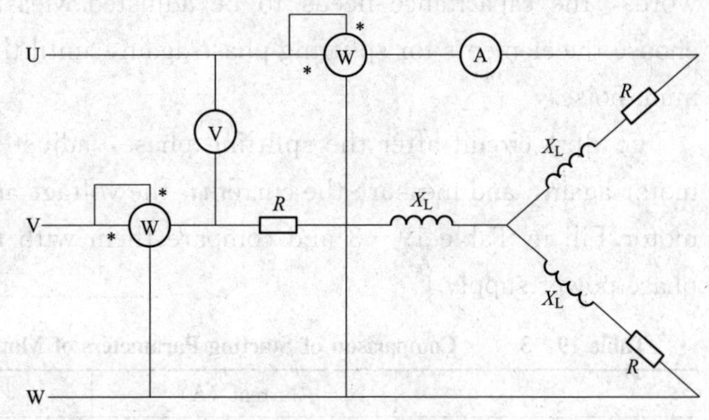

Figure 19 - 5　Measure the Starting Parameters of Three-Phase Asynchronous Motor Under the Supply of Symmetrical Three-Phase Power Supply

Table 19 - 2 the Single-Phase Parameters of Three-Phase Asynchronous Motor Starting (Locked Rotor) under the Supply of Symmetrical Three-Phase Power Supply

Measure	$I=0.6I_N$ (A)	U_{UV} (V)	P_{UW} (W)	P_{VW} (W)
Calculate	ΣP (W)			
	Per Phase Power P (W)			
	Per Phase Resistance R (Ω)			
	Per Phase Inductive Reactance X (Ω)			

In the table, the resistance $R=P/I^2$, $X=\sqrt{(U/I)^2-R^2}$ (U is the phase voltage).

3. Calculate the Capacitance (Inductance) Connected to the Motor Windings

Calculate the capacitive reactance (inductive reactance) of the capacitor (inductor) connect to the motor windings according to R and X in table 19 - 2.

If $X_L>\sqrt{3}R$, then

$$X_{C2}=X_L+\sqrt{3}R$$
$$X_{C3}=X_L-\sqrt{3}R$$

If $X_L<\sqrt{3}R$, then a compensating inductance shall be connected to one of the phases, there are

$$X_{C2}=X_L+\sqrt{3}R$$
$$X_{L3}=\sqrt{3}R-X_L$$

Calculate the capacitance or inductance according to $C=\frac{1}{\omega X_C}$ and $L_3=\frac{X_{L3}}{\omega}$, note down the calculated result.

4. Connect C_2 to phase B and C_3 (or L_3) to phase C of the motor. Add a single-phase power supply to the motor, and adjust the voltage regulator slowly. Observe whether the motor is rotating, and whether the rotation of the motor is stable if it is rotating. It shows that the three phases are not symmetrical enough if the rotating is not stable, in other words, the capacitance needs to be adjusted. Measure the parameters of the motor and choose the elements for splitting phase again, until the motor rotates stably and makes minimum noise.

5. In a circuit after the splitting phase, adjust the voltage regulator to zero. Start the motor again, and measure the current, the voltage and the power at the starting state of the motor. Fill in Table 19 - 3 and compare them with the values of the starting under three-phase power supply.

Table 19 - 3 Comparison of Starting Parameters of Motor under Two Kinds of Power Supply

	Current (A)	Voltage (V)	Power (W)
Three-Phase Power Supply			
Single-Phase Power Supply			
Difference			

6. Observe the starting of the motor until it rotates stably. Measure the input voltage and the phase voltage of the motor windings, and fill in Table 19 - 4.

Table 19 - 4 the Input Voltage of the Splitting Phase Circuit and the Output Voltage of the Three Phases

Input Voltage (V)	Output Voltage (V)		
	U_{AB}	U_{BC}	U_{CA}
40			
50			
60			

● **Notes**

1. When the single-phase, the two-phase and the three-phase power supplies are being compared, adjust the output of voltage regulator to zero and let the rotor be stationary each time in the starting state. Then increase the output of the voltage regulator and stop increasing voltage when the rotor begins to rotate (or the line voltage of power supply reaches 380V), write down the state.

2. At the locked rotor state, adjust the voltage regulator to zero first, and lock the rotor when it is stationary. Then increase the voltage slowly. No sudden addition of voltage on the motor whose rotor gets locked is allowed, and no sudden locking for a rotating rotor is allowed.

3. This experiment only verifies that the power supply can be transformed from single-phase to three-phase. The practice of this experiment must not be used in the actual operation of the motor.

● **Questions**

1. What is the difference between the starting parameters and the running parameters of a three-phase motor?

2. Why don't test the phase voltage directly when measuring the parameters of the motor? Why don't use three-wattmeter method to measure power?

● **Experiment Report**

1. Derive the calculation formula of the parameters of splitting phase circuit when $X_L < \sqrt{3}R$.

2. Analyze the results of the splitting phase experiment, and sum up the characteristics of the splitting phase circuit.

实验 20 负阻抗变换器

一、实验目的

(1) 加深对负阻抗概念的认识，掌握对含有负阻抗器件电路的分析方法。

(2) 了解负阻抗变换器的组成原理及其应用。

(3) 掌握负阻抗变换器的各种测试方法。

二、实验原理

负阻抗是电路理论中的一个重要的基本概念，在工程实践中也有广泛的应用。对于负阻抗的产生来说，除某些非线性元件（如隧道二极管）在某个电压或电流的范围内具有负阻抗特性外，一般都由一个有源双口网络来形成一个等值的线性负阻抗。该网络由线性集成电路或晶体管等元件组成，称作负阻抗变换器。

按有源网络输入电压和电流与输出电压和电流的关系，负阻抗变换器可分为电流倒置型（INIC）和电压倒置型（VNIC）两种，电路模型如图 20－1 所示。

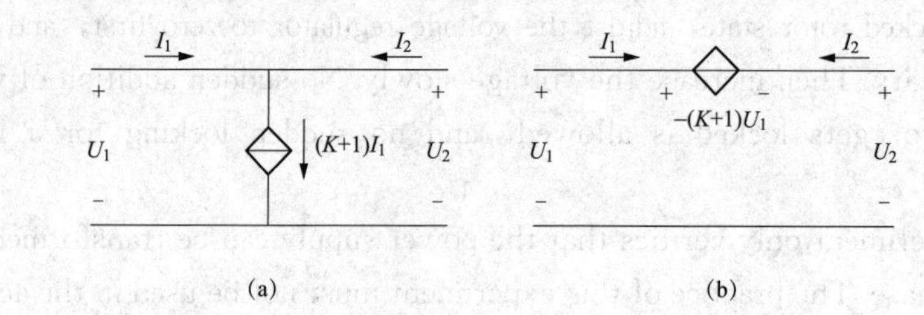

图 20－1 负阻抗变换器的两种电路模型

(a) 电流倒置型；(b) 电压倒置型

在理想情况下，其电压、电流关系如下。

对于 INIC 型：$U_2=U_1$，$I_2=K_1I_1$（K_1为电流增益）。

对于 VNIC 型：$U_2=-K_2U_1$，$I_2=-I_1$（K_2为电压增益）。

如果在 INIC 的输出端接上负载阻抗 Z_L，如图 20－2 所示，则它的输入阻抗 Z_i为

$$Z_i=\frac{U_1}{I_1}=\frac{U_2}{I_2/K_1}=\frac{K_1U_2}{I_2}=-K_1Z_L$$

即输入阻抗 Z_i为负载阻抗 Z_L的 K_1倍，且为负值，呈负阻特性。

本实验用线性运算放大器组成如图 20－3 所

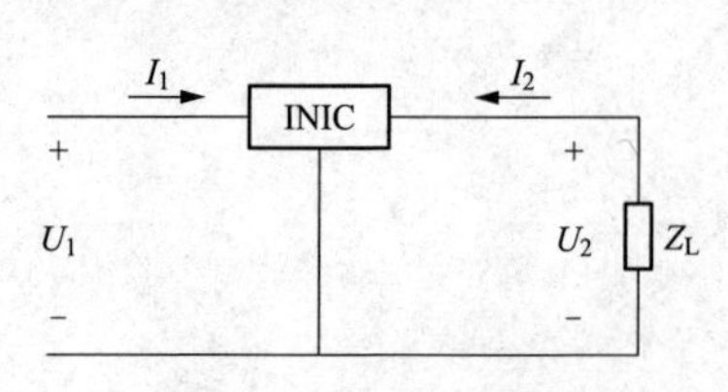

图 20－2 INIC 连接负载

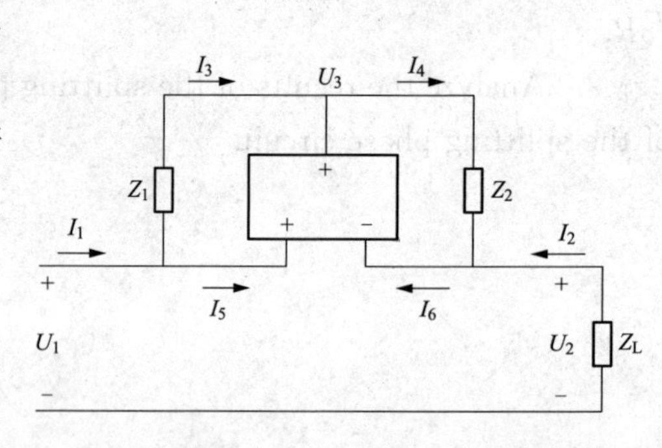

图 20－3 使用运算放大器构建负阻抗变换器

示的电路，在一定的电压、电流范围内可获得良好的线性度。

根据运放原理可知 $U_1=U_+=U_-=U_2$

又有 $I_5=I_6=0$，$I_1=I_3$，$I_2=-I_4$

所以

$$I_4Z_2=-I_3Z_1$$

$$-I_2Z_2=-I_3Z_1$$

$$\frac{U_2}{Z_L}Z_2=-I_1Z_1$$

$$\frac{U_2}{I_1}=\frac{U_1}{I_1}=Z_i=-\frac{Z_1}{Z_2}Z_L=-KZ_L$$

可见，该电路属于电流倒置型（INIC）负阻抗变换器，输入阻抗 Z_i 等于负载阻抗 Z_L 乘以 $-K$ 倍。

负阻抗变换器具有十分广泛的应用，例如可以用来实现阻抗变换。

假设 $Z_1=R_1=1\text{k}\Omega$，$Z_2=R_2=300\Omega$ 时

$$K=\frac{Z_1}{Z_2}=\frac{R_1}{R_2}=\frac{10}{3}$$

若负载为电阻，$Z_L=R_L$ 时

$$Z_1=-KZ_L=-\frac{10}{3}R_L$$

若负载为电容 C，$Z_L=\frac{1}{\text{j}\omega C}$ 时

$$Z_1=-KZ_L=-\frac{10}{3}\frac{1}{\text{j}\omega C}=\text{j}\omega L\left(令\ L=\frac{1}{\omega^2C}\times\frac{10}{3}\right)$$

若负载为电感 L，$Z_L=\text{j}\omega L$ 时

$$Z_1=-KZ_L=-\frac{10}{3}\text{j}\omega L=\frac{1}{\text{j}\omega C}\left(令\ C=\frac{1}{\omega^2L}\times\frac{3}{10}\right)$$

可见，电容通过负阻抗变换器呈现电感性质，而电感通过负阻抗变换器呈现电容性质。

三、实验设备

实验设备见表 20-1。

表 20-1　　实验设备

设备名称	型号与规格	数量	实验模块
恒压源	0～30V	1	NDG-02
恒流源	0～500mA	1	
直流电压表	0～200V	1	NDG-03
直流电流表	0～2000mA	1	
双踪示波器	GDS-1102A-U	1	
信号发生器	DG1022U	1	
实验电路	负阻抗变换器		NDG-11
可调电阻	0～9999Ω	1	NDG-06
电阻	510Ω	1	NDG-13

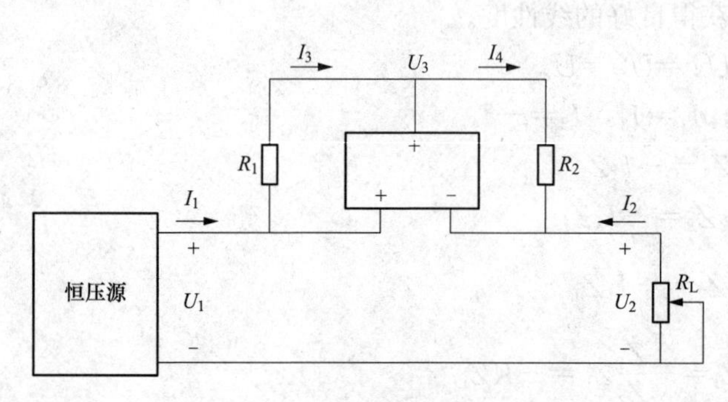

图 20-4　负阻抗变换器实验电路

四、实验内容

1. 测量负电阻的伏安特性

实验电路如图 20-4 所示，图中 U_1 为恒压源的输出电压，负载电阻 R_L 使用可调电阻。

（1）调节可调电阻，分别使 $R_L=300\Omega$ 及 600Ω。调节恒压源的输出电压，使之在（0～1V）范围内取值。分别测量 INIC 的输入电压 U_1 及输入电流 I_1，将数据记入表 20-2 中。

表 20-2　负电阻的伏安特性实验数据

$R_L=300\Omega$	U_1（V）	0.1	0.2	0.3	0.4	0.5	0.6	0.7	0.8	0.9	1
	I_1（mA）										
	$U_{1平均}$（V）					$I_{1平均}$（mA）					
$R_L=600\Omega$	U_1（V）	0.1	0.2	0.3	0.4	0.5	0.6	0.7	0.8	0.9	1
	I_1（mA）										
	$U_{1平均}$（V）					$I_{1平均}$（mA）					

（2）计算等效负阻。

实测值　$$R_-=\frac{U_{1平均}}{I_{1平均}}$$

理论计算值　$$R'_-=-KZ_L=-\frac{10}{3}R_L$$

电流增益　$$K=\frac{R_1}{R_2}$$

2. 阻抗变换及相位观察

用 0.1μF 的电容器（串联一 510Ω 电阻）和 10mH 的电感（串联一 510Ω 电阻）分别取代 R_L；用低频信号源（输出 $f=1\times10^3$ Hz 的正弦波形）取代恒压源。调节低频信号使 $U_1<1$V，用双踪示波器观察并记录 U_1 与 I_1 以及 U_2 与 I_2 的相位差（I_1、I_2 的波形分别从 R_1、R_2 两端取出）。

五、注意事项

（1）整个实验中应使 $U_1=(0\sim1)$V。

（2）防止运放输出端短路。

六、实验报告

（1）根据表 20-2 数据，完成要求的计算，并绘制负阻特性曲线 $U_1=f(I_1)$。

（2）根据实验内容 2 的数据，解释观察到的现象，说明负阻抗变换器实现阻抗变换的功能。

Experiment 20 Negative Impedance Converter

● Objectives

1. Acquire a better understanding of the concept of the negative impedance, and learn how to analyze a circuit with negative impedance.

2. Understand the composition principle of negative impedance converter and its application.

3. Learn the various testing methods for a negative impedance converter.

● Principles

Negative impedance is an important basic concept in circuit theory, and also widely used in engineering practice. In addition to the negative impedance of some nonlinear elements, an active two-port network is generally used to form the equivalent linear negative impedance. The network is called negative impedance converter, which is composed of elements such as linear integrated circuits or transistors.

According to the relationship between the input voltage and the input current, and the output voltage and the output current of the active network, the negative impedance converter can be divided into two kinds: a negative impedance converter with voltage inversion (VNIC) and a with current inversion (INIC). The two kinds of circuit models of the negative impedance converters is shown in Figure 20 - 1.

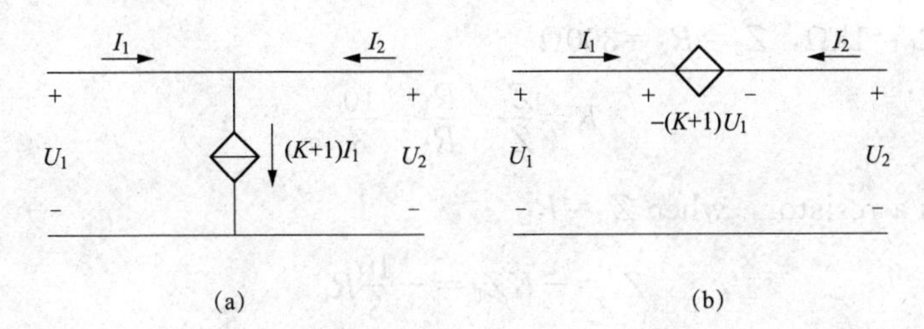

Figure 20 - 1 The Two Kinds of Circuit Models of Negative Impedance Converter

(a) VNIC; (b) INIC

In the ideal situation, the relationships between their voltages and current are:

for type INIC: $U_2=U_1$, $I_2=K_1I_1$ (K_1 is the current gain).

for type VNIC: $U_2=-K_2U_1$, $I_2=-I_1$ (K_2 is the voltage gain).

Connect the load impedance Z_L to the output port of INIC, as shown in Figure 20 - 2. The input impedance Z_i is

$$Z_i=\frac{U_1}{I_1}=\frac{U_2}{I_2/K_1}=\frac{K_1U_2}{I_2}=-K_1Z_L$$

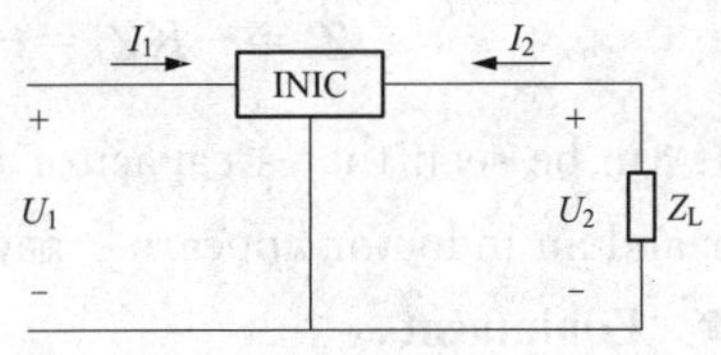

Figure 20 - 2 INIC Connected with a Load

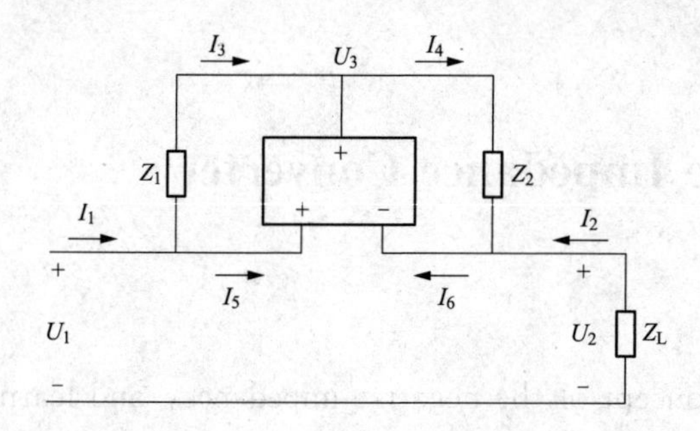

Figure 20 - 3 Negative Impedance Converter Circuit Constructed with Linear Operational Amplifier and Resistors

The input impedance Z_i is K_1 times the output impedance Z_L and negative, in other words, it appears negative impedance characteristics.

A negative impedance converter circuit constructed with a linear operational amplifier and some resistors is used in this experiment, as shown in Figure 20 - 3. Good linearity can be achieved within a certain range of voltage and current.

According to the principle of operational amplifier $U_1 = U_+ = U_- = U_2$ and $I_5 = I_6 = 0$, $I_1 = I_3$, $I_2 = -I_4$

$$I_4 Z_2 = -I_3 Z_1$$

$$-I_2 Z_2 = -I_3 Z_1$$

so

$$\frac{U_2}{Z_L} Z_2 = -I_1 Z_1$$

$$\frac{U_2}{I_1} = \frac{U_1}{I_1} = Z_i = -\frac{Z_1}{Z_2} Z_L = -K Z_L$$

It can be seen that this circuit is a negative impedance converter with current inversion (INIC), the input impedance Z_i is $-K$ times of the load impedance Z_L.

Negative impedance converters are applied widely. For example, it can be used to achieve impedance transformation.

When $Z_1 = R_1 = 1\text{k}\Omega$, $Z_2 = R_2 = 300\Omega$

$$K = \frac{Z_1}{Z_2} = \frac{R_1}{R_2} = \frac{10}{3}$$

If the load is a resistor, when $Z_L = R_L$

$$Z_1 = -K Z_L = -\frac{10}{3} R_L.$$

If the load is a capacitor C, when $Z_L = \frac{1}{j\omega C}$

$$Z_1 = -K Z_L = -\frac{10}{3}\frac{1}{j\omega C} = j\omega L\left(\text{letting } L = \frac{1}{\omega^2 C} \times \frac{10}{3}\right)$$

If the load is a inductor L, when $Z_L = j\omega L$,

$$Z_1 = -K Z_L = -\frac{10}{3} j\omega L = \frac{1}{j\omega C}\left(\text{letting } C = \frac{1}{\omega^2 L} \times \frac{3}{10}\right)$$

It can be seen that a capacitor appears inductance through the negative impedance converter and an inductor appears capacitance through the negative impedance converter.

- **Equipment**

Equipment is shown in Table 20 - 1.

Table 20-1 Equipment

Equipment	Model or Specification	Quantity	Module
Constant Voltage Source	0~30V	1	NDG-02
Constant Current Source	0~500mA	1	
DC Voltmeter	0~200V	1	NDG-03
DC Ammeter	0~2000mA	1	
Oscilloscope	GDS-1102A-U	1	
Signal Generator	DG1022U	1	
Experiment Circuit	Negative Impedance Converter		NDG-11
Adjustable Resistance	0~9999Ω	1	NDG-06
Resistor	510Ω	1	NDG-13

● Contents

1. Measure the Volt-Ampere Chara- cteristics of Negative Resistance

The experiment circuit is shown in Figure 20-4. U_1 is the output of the constant voltage source. The adjustable resistance is used for the load resistance R_L.

(1) Adjust the adjustable resistance, let $R_L=300\Omega$ and 600Ω respectively. Adjust the output of the constant voltage source, make the voltage values change in $0\sim1V$ range. Measure the input voltage U_1 and the input current I_1 of INIC, fill in Table 20-2.

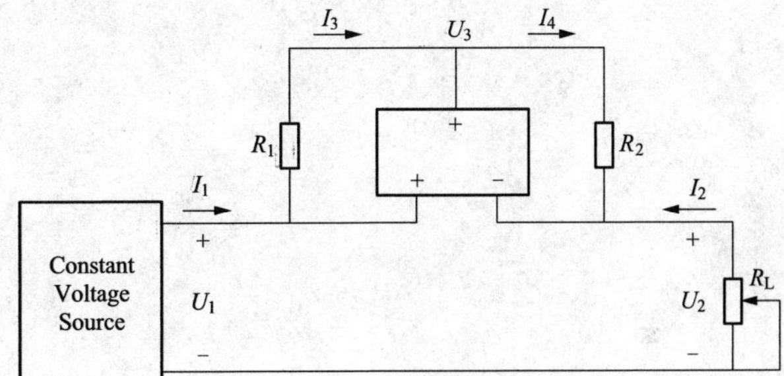

Figure 20-4 The Experiment Circuit of Negative Impedance Converter

Table 20-2 Experiment Data of Volt-Ampere Characteristics of Negative Resistance

$R_L=300\Omega$	U_1 (V)	0.1	0.2	0.3	0.4	0.5	0.6	0.7	0.8	0.9	1
	I_1 (mA)										
	$U_{1AVERAGE}$ (V)						$I_{1AVERAGE}$ (mA)				
$R_L=600\Omega$	U_1 (V)	0.1	0.2	0.3	0.4	0.5	0.6	0.7	0.8	0.9	1
	I_1 (mA)										
	$U_{1AVERAGE}$ (V)						$I_{1AVERAGE}$ (mA)				

(2) Calculate the Equivalent Negative Resistance

Measured value $\qquad R_-=\dfrac{U_{1AVERAGE}}{I_{1AVERAGE}}$

Calculated value $\qquad R'_-=-KZ_L=-\dfrac{10}{3}R_L$

Current gain $\qquad K=\dfrac{R_1}{R_2}$

2. Impedance Conversion and Phase Observation

Replace R_L with a capacitor of 0.1μF (in series with a 510Ω resistor) and a inductor of 10mH (in series with a 510Ω resistor) respectively. And replace the constant voltage source with a low-frequency signal source which outputs a $f=1\times10^3$ Hz sinusoidal wave. Adjust the low-frequency signal, make $U_1<1$V. Observe and write down the phase differences between U_1 and I_1, and U_2 and I_2 (the waveforms of I_1 and I_2 are from R_1 and R_2).

● **Notes**

1. During the entire process of the experiment, keep $U_1=(0\sim1)$V.

2. Be careful of the short circuit of the output of the operational amplifier.

● **Experiment Report**

1. Using the data in Table 20-2, finish the required calculation, and draw the characteristic curve $U_1=f(I_1)$ of the negative resistance.

2. Using the data noted in step 2, explain the observed phenomenon, and explain the impedance conversion function of the negative resistance converter.

实验 21　回转器特性测试

一、实验目的

（1）了解回转器的结构和基本特性。

（2）测量回转器的基本参数。

（3）了解回转器的应用。

二、实验原理

回转器是一种有源非互易的两端口网络元件，电路符号及其等值电路如图 21－1 所示。

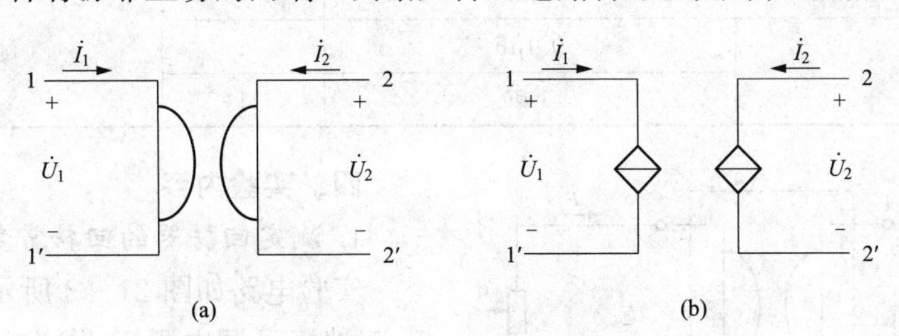

图 21－1　回转器

（a）回转器电路符号；（b）回转器等值电路

理想回转器的导纳方程为 $\begin{bmatrix}\dot{I}_1\\ \dot{I}_2\end{bmatrix}=\begin{bmatrix}0 & G\\ -G & 0\end{bmatrix}\begin{bmatrix}\dot{U}_1\\ \dot{U}_2\end{bmatrix}$

或写成　$\dot{I}_1=G\dot{U}_2\quad \dot{I}_2=-G\dot{U}_1$

也可写成电阻方程 $\begin{bmatrix}\dot{U}_1\\ \dot{U}_2\end{bmatrix}=\begin{bmatrix}0 & -R\\ +R & 0\end{bmatrix}\begin{bmatrix}\dot{I}_1\\ \dot{I}_2\end{bmatrix}$

或写成　$\dot{U}_1=R\dot{I}_2\quad \dot{U}_2=-R\dot{I}_1$

式中：G 和 R 分别称为回转电导和回转电阻，简称为回转常数。

若在 2－2′端接一负载电容 C，从 1－1′端看进去的导纳 Y_{i} 为

$$Y_{\mathrm{i}}=\frac{\dot{I}_1}{\dot{U}_1}=\frac{G\dot{U}_2}{-\dot{I}_2/G}=\frac{-G^2\dot{U}_2}{\dot{I}_2}$$

又因为

$$\frac{\dot{U}_2}{\dot{I}_2}=-Z_{\mathrm{L}}=-\frac{1}{\mathrm{j}\omega C}$$

所以

$$Y_{\mathrm{i}}=\frac{G^2}{\mathrm{j}\omega C}=\frac{1}{\mathrm{j}\omega L}$$

其中

$$L=\frac{C}{G^2}$$

可见，从 1－1′端看进去就相当于一个电感，即回转器能把一个电容元件“回转”成一个电感元件，所以也称为阻抗逆变器。由于回转器有阻抗逆变作用，在集成电路中得到重要的应用。因为在集成电路制造中，制造一个电容元件比制造电感元件容易得多，通常可以用一个带有电容负载的回转器来获得一个较大的电感负载。

三、实验设备

实验设备见表 21－1。

表 21－1　　**实 验 设 备**

设备名称	型号与规格	数量	实验模块
智能仪表	0～500V 0～3A	3	NDG－01
信号发生器	DG1022U	1	
双踪示波器	GDS－1102A－U	1	
实验电路	回转器		NDG－11
可调电阻	0～9999Ω	1	NDG－06
电阻	1kΩ	1	NDG－13
电容	0.1μF	1	
	1μF	1	

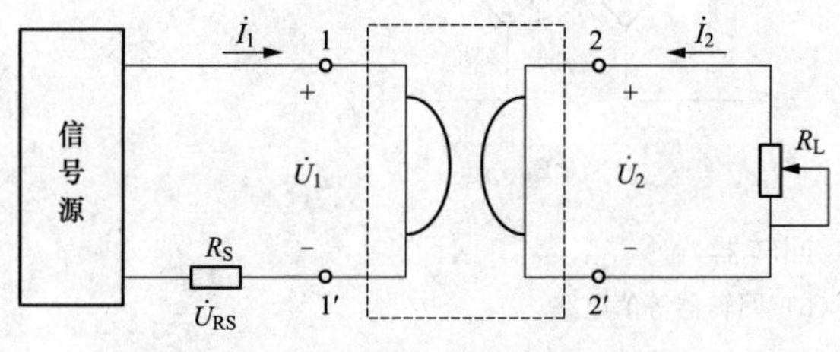

图 21－2　测定回转器的回转常数

四、实验内容

1. 测定回转器的回转常数

实验电路如图 21－2 所示，在回转器的 2－2′端接可调电阻 R_L 作为负载，取样电阻 R_S＝1kΩ，信号源频率固定在 1kHz，输出电压为 1～2V。用示波器测量不同负载电阻 R_L 时的 U_1、U_2 和 U_{RS}，并计算相应的电流 I_1、I_2 和回转常数 G，一并记入表 21－2 中。

表 21－2　　**测定回转常数的实验数据**

R_L (kΩ)	测量值		计算值				
	U_1 (V)	U_2 (V)	I_1 (mA)	I_2 (mA)	$G'=I_1/U_2$	$G''=I_2/U_1$	$G_{平均}=(G'+G'')/2$
0.5							
1							
1.5							
2							
3							
4							
5							

2. 测试回转器的阻抗逆变性质

（1）观察相位关系。实验电路如图 21－2 所示，在回转器 2－2′端的电阻负载 R_L 用电容 C 代替，且 C＝0.1μF。用双踪示波器观察回转器输入电压 U_1 和输入电流 I_1 之间的相位关系。图中 R_S 为电流取样电阻，因为电阻两端的电压波形与通过电阻的电流波形同相，所以用示波器观测电压 U_{RS} 的相位就相当于观测电流 I_1 的相位。

（2）测量等效电感。2－2′两端接负载电容 C＝0.1μF，用示波器测量不同频率时的等效电感，并算出 I_1、L'、L 及误差 ΔL，分析 U、U_1、U_{RS} 之间的相量关系。

3. 测量谐振特性

实验电路如图 21－3 所示，图中 $C_1=1\mu F$，$C_2=0.1\mu F$，取样电阻 $R_S=1k\Omega$。用回转器作电感，与 C_1 构成并联谐振电路。信号源输出电压保持恒定 $U=2V$，在不同频率时用示波器测量表 21－3 中规定的各个电压，并找出 U_1 的峰值。将测量数据和计算值记入表 21－3 中。

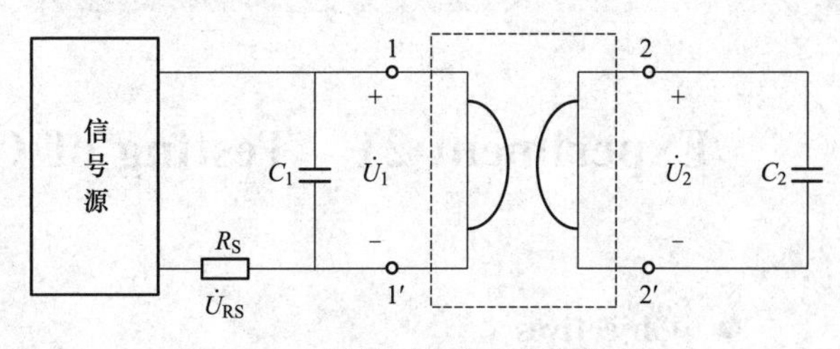

图 21－3　测量回转器的谐振特性

表 21－3　谐振特性实验数据

f（Hz）参数	200	400	500	700	800	900	1000	1200	1300	1500	2000
U_1（V）											
U_{RS}（V）											
$I_1=U_{RS}/R_S$（mA）											
$L'=U_1/2\pi f I_1$											
$L=C/G^2$											
$\Delta L=L'-L$											

五、注意事项

（1）回转器的正常工作条件是 U、I 的波形必须是正弦波，为避免运放进入饱和状态使波形失真，所以输入电压以不超过 2V 为宜。

（2）防止运放输出对地短路。

六、实验报告

（1）根据表 21－2 数据，计算回转电导。

（2）根据实验内容 2 的结果，画出电压、电流波形，说明回转器的阻抗逆变作用，并计算等效电感值。

（3）根据表 21－3 数据，画出并联谐振曲线，找到谐振频率，并和计算值相比较。

（4）从各实验结果中总结回转器的性质、特点和应用。

Experiment 21 Testing of Characteristics of Gyrator

- **Objectives**

1. Understand the structure and basic characteristics of gyrator.
2. Measure the basic parameters of the gyrator.
3. Know the application of gyrator.

- **Principles**

A gyrator is an active and nonreciprocal two-port circuit element, the circuit symbol and the equivalent circuit of a gyrator is shown in Figure 21 - 1.

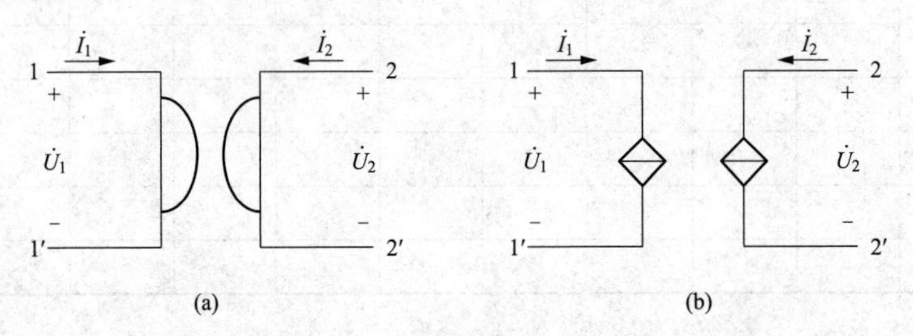

Figure 21 - 1 Gyrator

(a) The Circuit Symbol of Gyrator; (b) The Equivalent Circuit of Gyrator

The admittance equation of an ideal gyrator is

$$\begin{bmatrix}\dot{I}_1\\ \dot{I}_2\end{bmatrix}=\begin{bmatrix}0 & G\\ -G & 0\end{bmatrix}\begin{bmatrix}\dot{U}_1\\ \dot{U}_2\end{bmatrix}$$

Or $\dot{I}_1=G\dot{U}_2\,\dot{I}_2=-G\dot{U}_1$

It can also be written as a resistance equation

$$\begin{bmatrix}\dot{U}_1\\ \dot{U}_2\end{bmatrix}=\begin{bmatrix}0 & -R\\ +R & 0\end{bmatrix}\begin{bmatrix}\dot{I}_1\\ \dot{I}_2\end{bmatrix}$$

Or $\dot{U}_1=R\dot{I}_2\,\dot{U}_2=-R\dot{I}_1$

G and R in the equations are called gyration conductance and gyration resistance respectively, or for short, gyration constants.

If a load capacitor is connected to port 2 - 2′, the admittance Y_i seen from port 1 - 1′ is

$$Y_{\mathrm{i}}=\frac{\dot{I}_1}{\dot{U}_1}=\frac{G\dot{U}_2}{-\dot{I}_2/G}=\frac{-G^2\dot{U}_2}{\dot{I}_2}$$

And $$\frac{\dot{U}_2}{\dot{I}_2}=-Z_{\mathrm{L}}=-\frac{1}{\mathrm{j}\omega C}$$

so $$Y_{\mathrm{i}}=\frac{G^2}{\mathrm{j}\omega C}=\frac{1}{\mathrm{j}\omega L}$$

in the equation $$L=\frac{C}{G^2}$$

It can be seen that the element is equivalent to an inductor if seen from port 1 - 1′. In other words, the gyrator can "gyrate" a capacitor to an inductor, thus it is also called an impedance inverter. Due to the impedance inverting effect of the gyrator, it has been importantly used in integrated circuits. In the manufacturing of integrated circuits, it is much easier to make a capacitor than to make an inductor. A gyrator with a capacitance load can usually be used to obtain a larger inductance load.

● **Equipment**

Equipment is shown in Table 21 - 1.

Table 21 - 1 Equipment

Equipment	Model or Specification	Quantity	Module
Smart Meter	0~500V 0~3A	3	NDG - 01
Signal Generator	DG1022U	1	
Oscilloscope	GDS - 1102A - U	1	
Experiment Circuit	Gyrator		NDG - 11
Adjustable Resistance	0~9999Ω	1	NDG - 06
Resistor	1kΩ	1	NDG - 13
Capacitor	0. 1μF	1	
	1μF	1	

● **Contents**

1. Measure the Gyration Constant of the Gyrator

The experiment circuit is shown in Figure 21 - 2. Connect the adjustable resistance R_L to port 2 - 2′ of the gyrator as the load. The sampling resistor R_S = 1kΩ. Keep the frequency of the output of the signal generator at 1kHz. The output voltage of signal generator is 1~2V. Measure U_1, U_2 and U_{RS} under different load resistance R_L, and calculate the corresponding current I_1, I_2 and the gyration constant G, fill in Table 21 - 2.

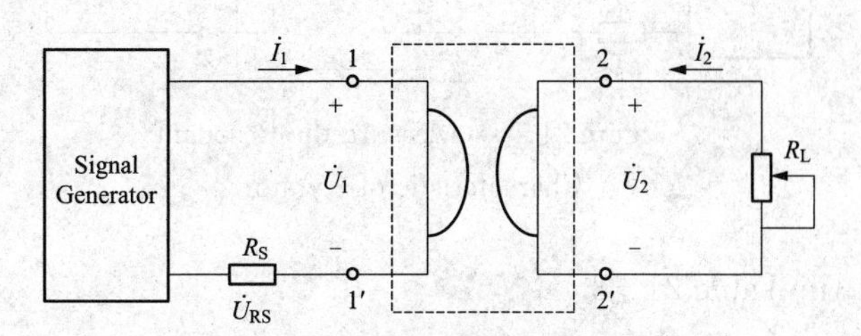

Figure 21 - 2 Measure the Gyration Constant of Gyrator

Table 21 - 2 Data of Measuring the Gyration Constant

R_L (kΩ)	Measured Value		Calculated Value				
	U_1(V)	U_2(V)	I_1(mA)	I_2(mA)	$G'=I_1/U_2$	$G''=I_2/U_1$	$G_{AVERAGE}=(G'+G'')/2$
0. 5							
1							
1. 5							
2							
3							
4							
5							

2. Test the Impedance Inverting Characteristic of the Gyrator

(1) Observe the Phase Relationship

The experiment circuit is shown in Figure 21 - 2. Replace the resistance load R_L connected to port 2 - 2′ of the gyrator with the capacitance C. Observe the phase relationship between the input voltage U_1 and the input current I_1 of the gyrator with oscilloscope. R_S in the figure is the current sampling resistor, because the waveform of the voltage across the resistor is in phase with the waveform of the current through the resistor. Observing the phase of voltage U_{RS} is the equal of observing the phase of current I_1.

(2) Measure the Equivalent Inductance

Connect the capacitor C = 0. 1μF to port 2 - 2′, and measure the equivalent inductances under different frequencies with oscilloscope. Calculate I_1, L', L and the error ΔL, and analyze the phase relationships between U, U_1 and U_{RS}.

3. Measure the Resonant Characteristic

The experiment circuit is shown in Figure 21 - 3. In the circuit: C_1=1μF, C_2=0. 1μF, and the sampling resistor R_S = 1kΩ. Use the gyrator as an inductor and form a parallel resonant circuit with C_1. Keep the output voltage of the signal generator unchanged (U = 2V), and measure the designated voltages in Table 21 - 3 under different frequencies with oscilloscope. Find the maximum value of U_1. And fill in Table 21 - 3.

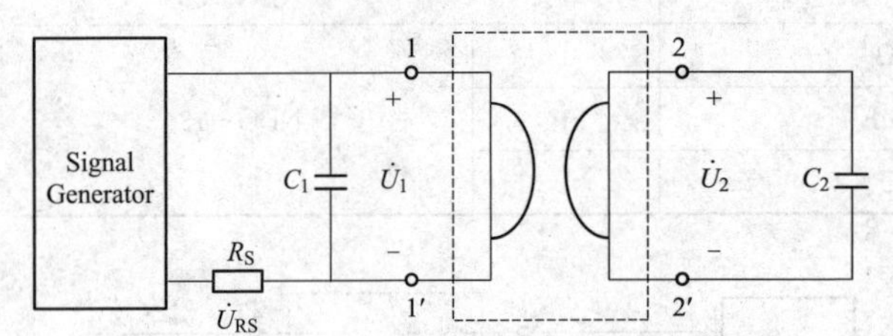

Figure 21 - 3 Measure the Resonant Characteristic of Gyrator

Table 21 - 3 Experiment Data of the Resonant Characteristic

Parameter \ f(Hz)	200	400	500	700	800	900	1000	1200	1300	1500	2000
U_1(V)											
U_{RS}(V)											
$I_1=U_{RS}/R_S$(mA)											
$L'=U_1/2\pi f I_1$											
$L=C/G^2$											
$\Delta L=L'-L$											

- **Notes**

1. The normal working conditions of a gyrator are the waveforms of U and I are sinusoidal waves. To prevent the operational amplifier from entering the saturation state and distorting the waveform, the input voltage shall not be more than 2V.

2. Beware of the operation amplifier's output shorts to ground.

● **Experiment Report**

1. Calculate the gyration conductance using the data in Table 21 - 2.

2. Using to the results of step 2, draw the waveforms of the voltage and the current. Explain the impedance inverting effect of the gyrator, and calculate the equivalent inductance.

3. Draw the parallel resonance curve, and find the resonant frequency and compare it with the calculated value.

4. Summarize the properties, characteristics and applications of gyrator using the results of the experiment.

实验 22　单相变压器特性测试

一、实验目的

（1）通过空载和短路实验测定变压器的变比和参数。

（2）通过负载实验测量变压器的运行特性。

二、实验原理

（1）铁芯变压器是一个非线性元件，铁芯中的磁感应强度 B 决定于外加电压的有效值 U。当二次侧开路（即空载）时，一次侧的励磁电流 I_O 与磁场强度 H 成正比。在变压器中，二次侧空载时，一次侧的电流和输入功率与所加电压之间的关系称为变压器的空载特性，这与铁芯的磁化曲线（$B-H$ 曲线）是一致的。空载实验通常是将高压侧开路，由低压侧通电进行测量。因变压器空载时阻抗很大，故电压表应接在电流表外侧。

空载时，一次侧功率、电压、电流为分别为 P_O、U_O、I_O，有

$$P_O = U_O I_O \cos\varphi_O$$

从空载特性曲线 $U_O = f(I_O)$，$P_O = f(U_O)$ 上查出对应于 $U_O = U_N$ 时的 I_O 和 P_O 值，可由下式算出励磁参数

$$r_m = \frac{P_O}{I_O^2}$$

$$Z_m = \frac{U_O}{I_O}$$

$$X_m = \sqrt{Z_m^2 - r_m^2}$$

（2）二次侧短路时，一次侧的电流和输入功率与所加电压之间的关系称为短路特性。变压器的短路阻抗很小，额定负载时的短路压降一般只有额定电压的 10% 左右。因此，短路实验所加电压不高。短路实验应将低压侧短路，在高压侧加电压进行测量。

短路时，一次侧功率、电压、电流为分别为 P_K、U_K、I_K，有

$$P_K = U_K I_K \cos\varphi_K$$

从短路特性曲线 $U_K = f(I_K)$，$P_K = f(I_K)$ 上查出对应于短路电流 $I_K = I_N$ 时的 U_K 和 P_K 值，可由下式算出短路参数

$$Z_K' = \frac{U_K}{I_K}$$

$$r_K' = \frac{P_K}{I_K^2}$$

$$X_K' = \sqrt{Z_K'^2 - r_K'^2}$$

折算到低压侧

$$Z_K = \frac{Z_K'}{K^2}$$

$$r_K = \frac{r_K'}{K^2}$$

$$X_K=\frac{X'_K}{K^2}$$

由于短路电阻 r_K 随温度而变化，因此，算出的短路电阻应按 GB/T 1032—2012《三相异步电动机试验方法》换算到基准工作温度 75℃时的阻值。

$$r_{K75℃}=r_{K\theta}\frac{234.5+75}{234.5+\theta}$$

$$Z_{K75℃}=\sqrt{r_{K75℃}+X_K^2}$$

式中：$r_{K\theta}$ 为室温下的短路电阻；234.5 为铜导线的常数，若用铝导线常数应改为 228；θ 为室温。

阻抗电压

$$U_K=\frac{I_N Z_{K75℃}}{U_N}\times 100\%$$

$$U_{Kr}=\frac{I_N r_{K75℃}}{U_N}\times 100\%$$

$$U_{KX}=\frac{I_N X_K}{U_N}\times 100\%$$

$I_K=I_N$ 时的短路损耗 $P_{KN}=I_N^2 r_{K75℃}$

(3) 变压器外特性测试。为了满足三组灯泡负载额定电压为 220V 的要求，故以变压器的低压（36V）绕组作为一次侧，220V 的高压绕组作为二次侧，即当作一台升压变压器使用。

在保持一次侧电压 $U_1=36V$ 不变时，逐次增加灯泡负载（每只灯为 25W），测定 U_1、U_2、I_1 和 I_2，即可绘出变压器的外特性，即负载特性曲线 $U_2=f(I_2)$。

三、实验设备

实验设备见表 22-1。

表 22-1　实验设备

设备名称	型号与规格	数量	实验模块
智能仪表	0～500V 0～3A	3	NDG-01
交流电源	0～450V 三相/0～250V 单相		QS-DYD3
白炽灯泡	25W	9	NDG-10
变压器	220V/36V	1	NDG-09

四、实验内容

1. 空载实验

实验线路如图 22-1。实验时，变压器低压线圈 a、x 接调压器输出，高压线圈 A、X 开路。

(1) 合闸前，将调压器置零，并合理选择仪表量程。

(2) 将电源合闸，缓慢调节调压器，直至变压器空载电压 $U_O=1.2U_N$。

(3) 然后，在 $1.2U_N$～$0.5U_N$ 的范围内逐次降低电源电压，测取变压器的 U_O、I_O、P_O，共取 6～7 组数据，记录于表 22-2 中。其中 $U=U_N$ 的点必须测，并在该点附近多取数点。

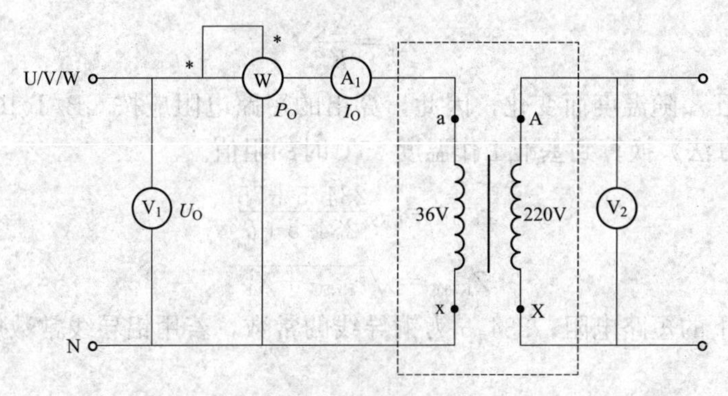

图 22-1　变压器空载实验

为了计算变压器的变化，在U_N以下测取一次侧电压的同时测取二次侧电压，填入表 22-2 中。测量数据以后，将调压器回零。

表 22-2　　变压器空载试验数据

序号	实验数据				计算数据
1	U_O（V）	I_O（A）	P_O（W）	U_{AX}	$\cos\varphi_2$
2					
3					
4					
5					
6					
7					

2. 短路实验

实验线路如图 22-2。

注意：每次改接线路时，都要将调压器回零。

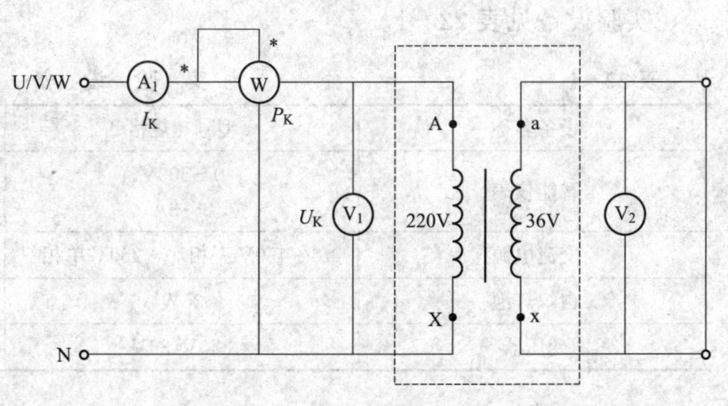

图 22-2　变压器短路实验

实验时，变压器的高压线圈接电源，低压线圈直接短路。接通交流电源，逐次增加输入电压，直到短路电流等于 $1.1I_N$为止。在 $0.5I_N$～$1.1I_N$范围内测取变压器的U_K、I_K、P_K，共取 6～7 组数据记录于表 22-3 中，其中 $I=I_N$的点必测。并记录实验时周围环境温度（℃）。

表 22-3　　变压器短路实验数据　　$\theta=$　　℃

序号	实验数据			计算数据
	U（V）	I（A）	P（W）	$\cos\varphi_k$
1				
2				

续表

序号	实验数据			计算数据
	U（V）	I（A）	P（W）	$\cos\varphi_k$
3				
4				
5				
6				

3. 负载实验

实验线路如图 22－3 所示。变压器低压线圈接电源，高压线圈接灯泡负载。加电源前将调压器回零，所有灯泡开关断开。

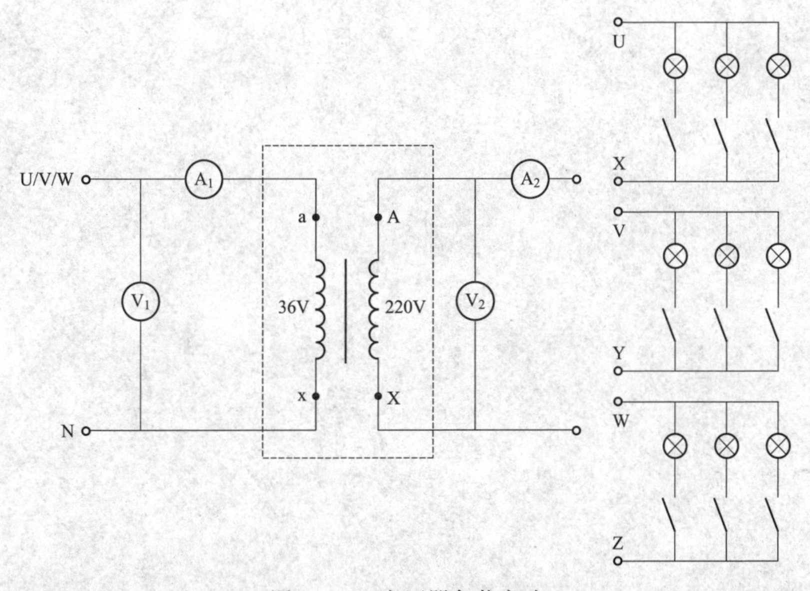

图 22－3　变压器负载实验

将灯泡负载连接后，合上交流主电源，逐渐升高调压器电压，使变压器输入电压 $U_1 = U_N = 36V$，测取变压器的输出电压 U_2 和电流 I_2。按以下要求测量数据，填入表 22－4 中：①负载为三个灯泡串联；②负载为两个灯泡串联；③负载为两个灯泡相并联再与一个灯泡相串联；④负载为一个灯泡；⑤负载为两个灯泡并联。

以上每次测量结束后，均将调压器回零，关闭电源开关，改接电路后重新打开电源，加压至指定值，再进行测量。实验完成后，断开三相交流电源，并将调压器回零。

表 22－4　　变压器负载实验数据

序号	1	2	3	4	5
U_2（V）					
I_2（A）					

五、注意事项

短路实验操作要快，否则线圈发热会引起电阻变化。

六、实验报告

（1）选取空载实验中测得变压器的一次侧、二次侧电压的三组数据，分别计算出变比，然后取其平均值作为变压器的变比 K。

（2）绘制出空载特性曲线 $U_O=f(I_O)$、$P_O=f(U_O)$，并计算出励磁参数。

（3）绘制出短路特性曲线 $U_K=f(I_K)$，$P_K=f(I_K)$，并计算出短路参数。

（4）绘制出负载特性曲线 $U_2=f(I_2)$。

（5）利用空载和短路实验测定的参数，画出被测变压器折算到低压侧的 Γ 型等效电路。

Experiment 22 Testing of Characteristics of Single-Phase Transformer

● **Objectives**

1. Measure the transformation ratio and parameters of the transformer by the no-load and the short-circuit experiment.

2. Measure the operation characteristics of the transformer by the load experiment.

● **Principles**

1. An iron core transformer is a nonlinear element. The magnetic induction B in the iron core is determined by the voltage RMS U of the applied voltage. When the secondary side of the transformer is open (no load), the field current I_O of the primary side is proportional to the magnetic field intensity H. In a transformer, when the secondary side is no load, the relationship between the current of the primary side and the applied voltage and the relationship between the input power of the primary side and the applied voltage are called the no-load characteristics of the transformer. In a no-load experiment, the high voltage side is open and the voltage is added to the low voltage side. The voltmeter should be connected to the outside of the ammeter because the no-load resistance of the transformer is high.

When the transformer is no load, the power, voltage and current of the primary side are P_O, U_O and I_O, there is

$$P_O = U_O I_O \cos\varphi_O$$

Find the value of I_O and P_O corresponding to $U_O = U_N$ on the no-load characteristic curves $U_O = f(I_O)$ and $P_O = f(U_O)$, the excitation parameters can be calculated by the following

$$r_m = \frac{P_O}{I_O^2}$$

$$Z_m = \frac{U_O}{I_O}$$

$$X_m = \sqrt{Z_m^2 - r_m^2}$$

2. When the secondary side of the transformer is short circuited, the relationship between the current in the primary side and the applied voltage and the relation between the input power of the primary side and the applied voltage are called the short-circuit characteristics. The short-circuit resistance of the transformer is very low. The short-circuit voltage drop at the rated load is only about ten percent of the rated voltage. So the voltage of short circuit experiment is not high. In a short circuit experiment, the low voltage side is short circuited and the voltage is added to the high voltage side.

When the transformer is short circuited, the power, voltage and current of the primary side are P_K, U_K and I_K, there is

$$P_K = U_K I_K \cos\varphi_K$$

Find the value of U_K and P_K corresponding to $I_K = I_N$ on the short-circuit characteristic curves $U_K = f(I_K)$ and $P_O = f(I_K)$, the excitation parameters can be calculated by the following

$$Z'_K = \frac{U_K}{I_K}$$

$$r'_K = \frac{P_K}{I_K^2}$$

$$X'_K = \sqrt{Z_K'^2 - r_K'^2}$$

Convert to the low voltage side

$$Z_K = \frac{Z'_K}{K^2}$$

$$r_K = \frac{r'_K}{K^2}$$

$$X_K = \frac{X'_K}{K^2}$$

Because the short-circuit resistance r_K changes with the temperature, the calculated short-circuit resistance shall be converted to the resistance at the reference working temperature 75℃ according to the national standard GB/T 1032—2012 *Test Procedures for three-phase induction motors*

$$r_{K75℃} = r_{K\theta} \frac{234.5 + 75}{234.5 + \theta}$$

$$Z_{K75℃} = \sqrt{r_{K75℃} + X_K^2}$$

$r_{K\theta}$ is the short-circuit resistance under the room temperature. 234.5 is the constant of the copper wire. If the aluminium wire is used, the constant is 228. θ is the room temperature.

The impedance voltages

$$U_K = \frac{I_N Z_{K75℃}}{U_N} \times 100\%$$

$$U_{Kr} = \frac{I_N r_{K75℃}}{U_N} \times 100\%$$

$$U_{KX} = \frac{I_N X_K}{U_N} \times 100\%$$

The short-circuit loss when $I_K = I_N$: $P_{KN} = I_N^2 r_{K75℃}$

3. The External Characteristic Testing of the Transformer

In order to meet the requirements of the 220V rated voltage of three sets of bulbs, the low voltage (36V) winding of the transformer is used as the primary side and the high voltage (220V) winding is used as the secondary side. In other words, the transformer is used as a step-up transformer in the external characteristic testing experiment.

Keep the voltage on the primary side $U_1 = 36$V unchanged. Increase the bulb load (25W each bulb) gradually, and measure U_1, U_2, I_1 and I_2. The external characteristic of the transformer can be drawn, that is, the load characteristic curve $U_2=f(I_2)$.

● Equipment

Equipment is shown in Table 22 - 1.

Table 22 - 1 Equipment

Equipment	Model or Specification	Quantity	Module
Smart Meter	0～500V 0～3A	3	NDG - 01
AC Power Supply	0～450V Three-Phase 0～250V Single-Phase		QS - DYD3
Incandescent Bulb	25W	9	NDG - 10
Transformer	220V/36V	1	NDG - 09

● Contents

1. No Load Experiment

The experiment circuit is shown in Figure 22 - 1. The low voltage winding a-x is connected to the output of the voltage regulator and the high voltage winding A-X is open.

(1) Before the power switch is closed, adjust the voltage regulator to zero and choose the ranges of the instruments reasonably.

(2) Close the power switch and adjust the voltage regulator slowly until the no-load voltage of the transformer $U_O=1.2U_N$.

(3) Reduce the voltage of the power supply in the range of $1.2U_N$～$0.5U_N$ gradually. Measure U_O, I_O and P_O of the transformer and take a total seven sets of data, and fill in Table 22 - 2. The point $U=U_N$ must be measured, and more points near this point should be taken. In order to calculate the change of the transformer, the secondary side voltage should be measured when the primary side voltage is measured below U_N, fill in Table 22 - 2.

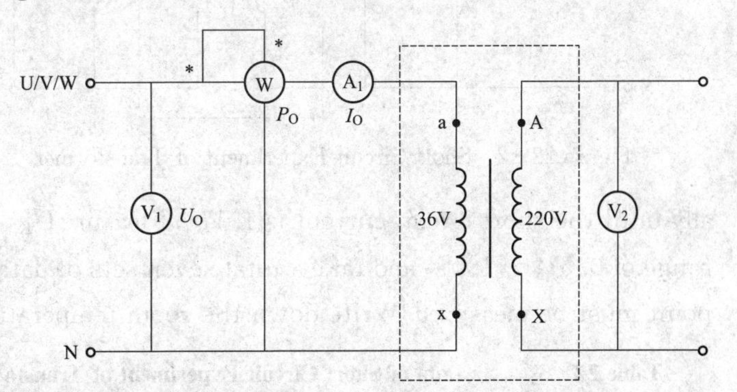

Figure 22 - 1 No Load Experiment of Transformer

(4) After the measurement, adjust the voltage regulator to zero for the next step of the experiment.

Table 22 - 2 Data of No Load Experiment of Transformer

Number	Measured Data				Calculated Data
	U_O (V)	I_O (A)	P_O (W)	U_{AX}	$\cos\varphi_2$
1					
2					
3					

续表

Number	Measured Data				Calculated Data
	U_O (V)	I_O (A)	P_O (W)	U_{AX}	$\cos\varphi_2$
4					
5					
6					
7					

2. Short Circuit Experiment

The experiment circuit is shown in Figure 22 - 2.

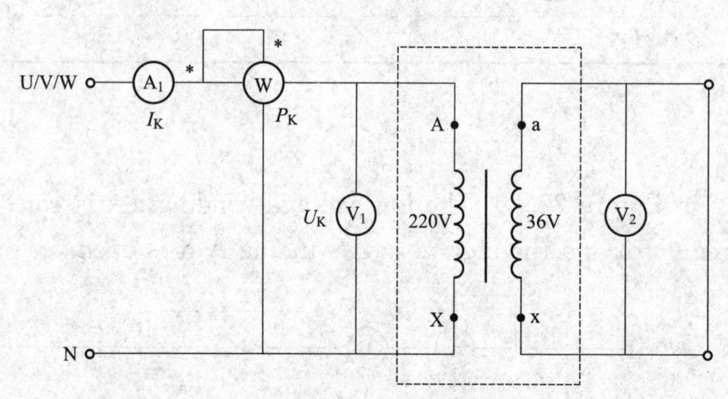

Figure 22 - 2 Short Circuit Experiment of Transformer

Note: every time the circuit is changed, the voltage regulator should be adjusted to zero.

During the process of the experiment, the high voltage winding of the transformer is connected to the power supply and the low voltage winding is short circuited directly. Close the power switch and increase the input voltage gradually until the short circuit current is $1.1I_N$. Measure U_K, I_K and P_K of the transformer in the range of $0.5I_N \sim 1.1I_N$ and take a total seven sets of data, and fill in Table 22 - 3. The $I=I_N$ point must be measured. Write down the room temperature (℃).

Table 22 - 3 Data of Short Circuit Experiment of Transformer Room Temperature $\theta=$ ℃

Number	Measured Data			Calculated Data
	U(V)	I(A)	P(W)	$\cos\varphi_k$
1				
2				
3				
4				
5				
6				
7				

3. Load Experiment

The experiment circuit is shown in Figure 22 - 3. The low voltage winding of the transformer is connected to the power supply and the high voltage winding is connected to the bulbs load. Before the power switch is closed, adjust the voltage regulator to zero and disconnect all the bulb switches.

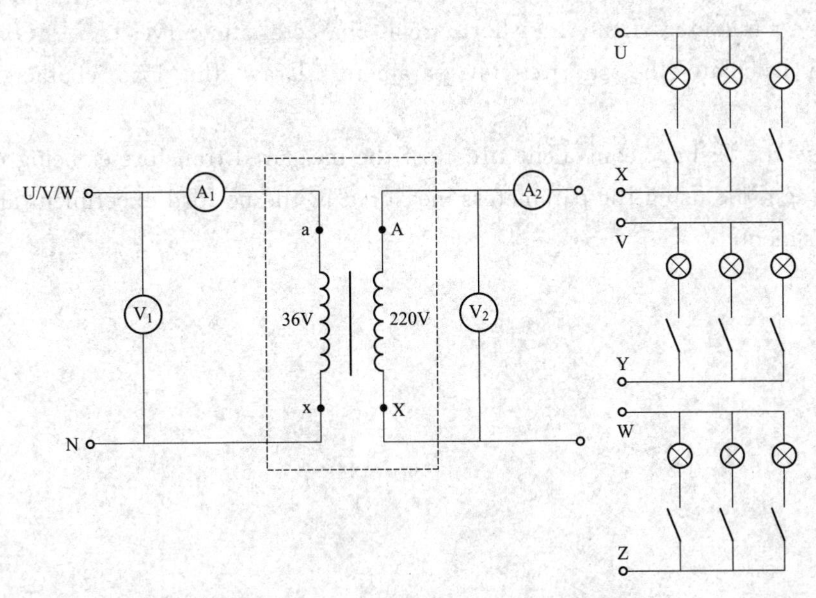

Figure 22 - 3　Load Experiment of Transformer

Connect the light bulb load to the transformer, and close the AC power switch. Increase the output of the voltage regulator until the input voltage of the transformer $U_1 = U_N = 36V$. Measure the output voltage U_2 and the output current I_2 of the transformer according to the following requirements, and fill in Table 22 - 4: ①The load is three bulbs in series. ②The load is two bulbs in series. ③The load is two bulbs in parallel and in series with one bulb. ④The load is one bulb. ⑤The load is two bulbs in parallel.

At the end of each measurement, adjust the voltage regulator to zero, and disconnect the power supply switch. Change the circuit connection and close the power supply switch and adjust the voltage to the designated value, and measure the data again. When the experiment is completed, disconnect the three-phase AC power supply, and adjust the voltage regulator to zero.

Table22 - 4　　Data of Load Experiment of Transformer

Number	1	2	3	4	5
U_2(V)					
I_2(A)					

- **Notes**

Short circuit experiment shall be operated quickly otherwise the heat of the coil will cause the resistance change.

- **Experiment Report**

1. Choose three sets of measured data of the voltages of the primary and secondary sides of the transformer from the no load experiment and calculate the transformation ratio respectively. Take the average value as the transformation ratio K of the transformer.

2. Draw the no-load characteristic curves $U_O = f(I_O)$ and $P_O = f(U_O)$, and calculate

the excitation parameters. Draw the short-circuit characteristic curves $U_K=f(I_K)$ and $P_K=f(I_K)$, and calculate the short-circuit parameters. Draw the load characteristic curve $U_2=f(I_2)$.

3. Draw the Γ-shape equivalent circuit of the measured transformer being converted to the low voltage side using the parameters measured in the no load experiment and the short circuit experiment.

实验 23　三相异步电动机点动与自锁控制

一、实验目的

(1) 熟悉三相笼型异步电动机点动和自锁控制线路中各元器件的使用方法及其在线路中所起的作用。

(2) 掌握三相笼型异步电动机点动和自锁控制线路的工作原理、接线方法、调试及故障排除技能。

二、实验原理

三相笼型异步电动机由于结构简单、性价比高、维修方便等优点获得了广泛的应用。在工农业生产中，经常采用继电器接触控制系统对中小功率笼型异步电机进行单向控制，其控制线路大部分由继电器、接触器、按钮等有触头电器组成。

如图 23-1 所示为三相笼型异步电动机点动与自锁控制线路。

启动时，合上漏电保护断路器和空气断路器 QF，引入三相电源。按下启动按钮 SB3 时，接触器 KM1 的线圈通电，主触头 KM1 闭合，电动机接通电源启动。当手松开按钮时，接触器 KM1 断电释放，主触头 KM1 断开，电动机电源被切断而停止运转。

当按下启动按钮 SB2 时，接触器 KM1 的线圈通电，主触头闭合，电动机接通电源启动。同时与 SB3 相连的接触器辅助动合触点 KM1 闭合并形成自锁。当手松开按钮时，由于辅助动合触点 KM1 闭合并自锁，所以电动机一直运转。要使电机停止运转，按下开关 SB1 即可。

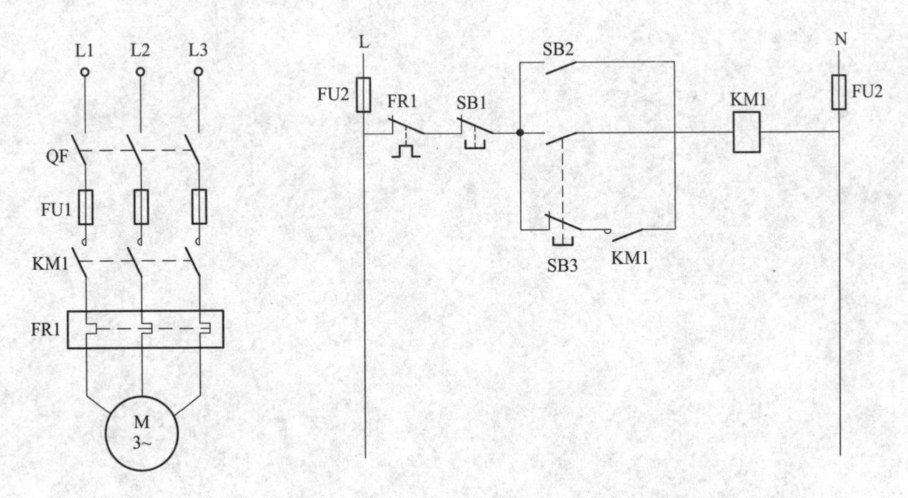

图 23-1　三相异步电动机点动与自锁控制线路

三、实验设备

实验设备见表 23-1。

表 23-1 实 验 设 备

设备名称	型号与规格	数量	实验模块
交流电源	0～450V 三相/0～250V 单相		QS-DYD3
三相笼型异步电动机	220V	1	M14
继电接触控制组件		1	EEL-57A

四、实验内容

(1) 检查各实验设备外观及质量是否良好。

(2) 按图 23-1 三相笼型异步电动机点动和自锁控制线路进行正确的接线。先接主回路，再接控制回路。自己检查无误并经指导老师检认可后方可合闸通电实验。

(3) 进行点动和连续运行操作。

1) 热继电器值调到 1.0A。

2) 合上漏电保护断路器和空气断路器 QF，引入三相电源。

3) 按下启动按钮 SB3，观察电机工作情况。

4) 按下按钮开关 SB2，观察电机工作情况。

5) 按下停止按钮 SB1，切断电机控制电源。

6) 断开空气断路器 QF，切断三相主电源。

7) 断开漏电保护断路器，关断总电源。

Experiment 23 Jog and Self-Locking Control of Three-Phase Asynchronous Motor

- **Objectives**

1. Know the use and the functions of the elements in the jog and self-locking control circuit of three-phase squirrel-cage asynchronous motor.

2. Learn the working principle, the connecting method, the testing and the trouble-shooting of the jog and self-locking control circuit of three-phase squirrel-cage asynchronous motor.

- **Principles**

Three-phase cage asynchronous motor has been widely used for its advantages of simple structure, high cost performance, convenient maintenance and so on. In the process of industrial and agricultural production, the contact control system of relay is often used to control the small and medium power cage asynchronous motor. Most of the control circuits are composed of relay, contactor, button and other contact electrical appliances.

The jog and self-locking control circuit of three-phase squirrel-cage asynchronous motor is shown in Figure 23 - 1.

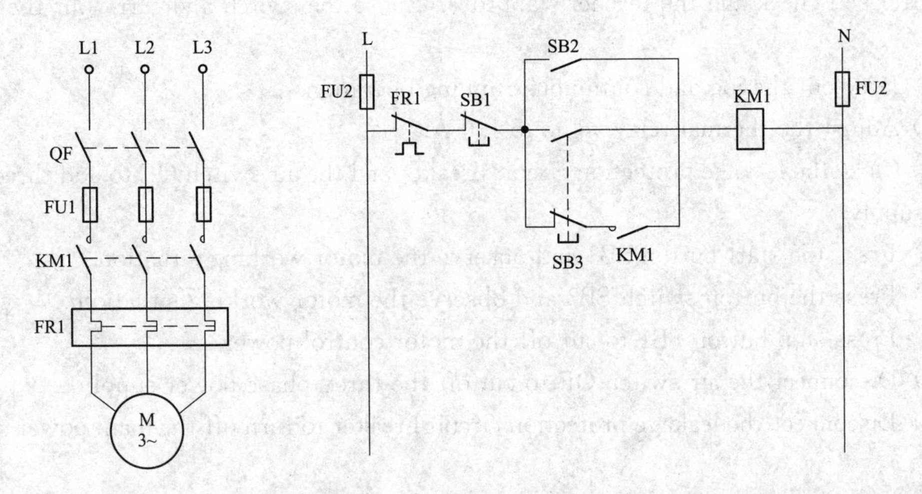

Figure 23 - 1 The Jog and Self-Locking Control Circuit of Three-Phase Squirrel-Cage Asynchronous Motor

When starting, close the leakage protection circuit breaker and the air switch QF to add the three-phase power supply. When the start button SB3 is pressed, the coil of the contactor KM1 is electrified, and the main contact KM1 is closed. The motor is turned on and starts running. When the button is loosened, the contact KM1 released and powered off, and the main contact KM1 is disconnected. The motor power is cut off and the motor stops running.

When the start button SB2 is pressed, the coil of the contactor KM1 is electrified, and the main contact is closed. The motor is turned on and starts running. At the same time, the auxiliary normally open contact KM1 connected to SB3 is closed and self-locked. When the button is loosened, because the auxiliary contact KM1 is closed and self-locked, the motor keeps running. To stop the motor from running, press the switch SB1.

- **Equipment**

Equipment is shown in Table 23 - 1.

Table 23 - 1 Equipment

Equipment	Model or Specification	Quantity	Module
AC Power Supply	0~450V Three-Phase 0~250V Single-Phase		QS - DYD3
Three-Phase Squirrel-Cage Asynchronous Motor	220V	1	M14
Relay Contact Control		1	EEL - 57A

- **Contents**

1. Check the appearance and quality of the experiment equipment.

2. Connect the jog and self-locking control circuit of three -phase squirrel-cage asynchronous motor shown in Figure 23 - 1. Connect the main loop first, then the control loop. After self-check and the teacher's approval, close the switch and carry out the experiment.

3. Carry out the jog and continuous running operation.

(1) Adjust the thermal relay value to 1. 0 A.

(2) Close the leakage protection circuit breaker and the air switch QF to add three-phase power supply.

(3) Press the start button SB3 and observe the motor working situation.

(4) Press the button switch SB3 and observe the motor working situation.

(5) Press stop button SB1 to cut off the motor control power.

(6) Disconnect the air switch QF to cut off the three-phase power supply

(7) Disconnect the leakage protection circuit breaker to turn off the main power supply.

实验 24　三相异步电动机正反转的控制

一、实验目的

（1）掌握三相笼型异步电动机正反转的工作原理、接线方式及操作方法。

（2）掌握机械及电气互锁的连接方法及其在控制线路中所起的作用。

（3）掌握按钮和接触器双重互锁控制的三相异步电动机正反转的控制线路。

二、实验原理

生产过程中，生产机械的运动部件往往要求能进行正反方向的运动，这就要求拖动电动机能作正反向旋转。由电机原理可知，将接至电机的三相电源进线中的任意两相对调，即可改变电动机的旋转方向。但为了避免误动作引起电源相间短路，往往在这两个相反方向的单相运行线路中加设必要的机械及电气互锁。按照电动机正反转操作顺序的不同，分别有“正—停—反”和“正—反—停”两种控制线路。对于“正—停—反”控制线路，要实现电动机有“正转—反转”或“反转—正转”的控制，都必须按下停止按钮，再向另一方向启动。然而对于生产过程中要求频繁的实现正反转的电机，为提高生产效率，减少辅助工时，往往要求能直接实现电动机正反转控制。

如图 24－1 所示为接触器和按钮双重连锁的三相异步电动机正反转控制线路。

启动时，合上漏电断路器及空气断路器 QF，引入三相电源。按下启动按钮 SB2，接触器 KM1 的线圈通电，主触头 KM1 闭合，且线圈 KM1 通过与开关 SB2 动合触点并联的辅助动合触点 KM1 实现自锁，同时通过按钮和接触器形成双重互锁。电动机正转运行。当按下按钮开关 SB3 时，接触器 KM2 的线圈通电，其主触头 KM2 闭合，且线圈 KM2 通过与开关 SB3 的动合触点并联的辅助动合触点 KM2 实现自锁。同时与接触器 KM1 互锁的动断触点都断开，使接触器 KM1 断电释放。电动机反转运行。要使电动机停止运行，按下开关 SB1 即可。

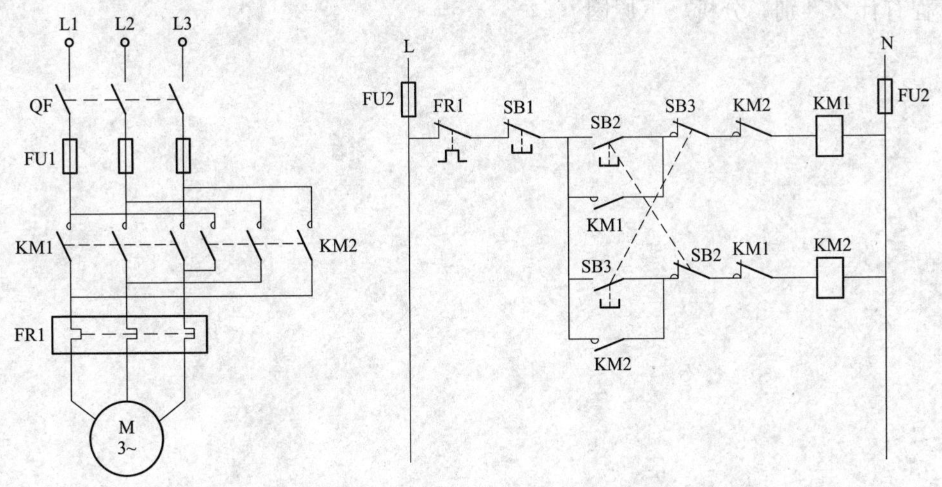

图 24－1　三相异步电动机正反转控制线路

三、实验设备

实验设备见表 24-1。

表 24-1 实 验 设 备

设备名称	型号与规格	数量	实验模块
交流电源	0～450V 三相/0～250V 单相		QS-DYD3
三相笼型异步电动机	220V	1	M14
继电接触控制组件		1	EEL-57A

四、实验内容

（1）检查各实验设备外观及质量是否良好。

（2）按图 24-1 三相笼型异步电动机接触器和按钮开关双重互锁控制正反转控制线路进行正确接线，先接主回路，再接控制回路。自己检查无误并经指导老师检查认可方可合闸实验。

（3）进行“正—反—停”操作。

1）热继电器值调到 1.0A。

2）合上漏电断路器及空气断路器 QF，引入三相电源。

3）按下按钮 SB2，观察电动机及各接触器的工作情况。

4）按下按钮 SB3，观察电动机的工作情况。

5）按下停止按钮 SB1，断开电机控制电源。

6）断开空气断路器 QF，切断三相主电源。

7）断开漏电保护断路器，关断总电源。

五、思考题

（1）在图 24-1 中，接触器和按钮是如何实现双重互锁的？

（2）双重互锁比起单重互锁的好处是什么？

（3）为什么要实现双重互锁？其意义何在？

（4）在上述实验当中，观察一下电动机在转换的过程中会出现什么情况？与“正—停—反”过程有什么区别，分析一下原因。

Experiment 24 Positive and Reverse Rotating Control of Three-Phase Asynchronous Motor

- **Objectives**

1. Understand the working principle, circuit connection and operation method of positive and reverse rotating of three-phase squirrel-case asynchronous motor.

2. Understand the connections of mechanical interlocking and electrical interlocking and their functions in the control circuit.

3. Learn the positive and reverse rotating control of the three-phase asynchronous motor with double interlocking control of buttons and contactors.

- **Principles**

In the process of industrial production, the moving parts of the production machines are often required to move in a positive and reverse direction, which requires that the drag motor can rotate positively and reversely. It is known from the principle of motor that swapping any two phases of the three-phase power supply wires connected to the motor can change the rotation direction of the motor. But in order to avoid that the malfunction causes short circuit of the power supply, necessary mechanical and electrical interlocking is often added to the two single phasing circuits in opposite directions. According to the difference in the sequence of positive and reverse operation of the motor, there are two kinds of control circuits, *positive-stop-reverse* and *positive-reverse-stop*. In the *positive-stop-reverse* control circuit, to achieve the *positive-reverse* or the *reverse-positive* control, the stop button must be pressed first then start in another direction. For the motors that requires frequent positive and reverse rotating in the process of production, to improve production efficiency and reduce auxiliary working hours, it is often required to bring about positive and reverse rotating control of motor directly.

The positive and reverse rotating control circuit of the three-phase asynchronous motor with double interlocking control of buttons and contactors is shown in Figure 24 - 1.

When starting, close the leakage protection circuit breaker and the air switch QF to add the three-phase power supply. Press the start button SB2, the coil of the contactor KM1 is electrified, and the main contact KM1 is closed. The coil KM1 is self-locked by the auxiliary often open contact KM1 in parallel with the often open contact of switch SB2. At the same time, double interlock is formed by the button and contactor. The motor rotates positively. When the button switch SB3 is pressed, the coil of the contact KM2 is electrified, and the main contact KM2 is closed. The coil KM2 is self-locked by the auxiliary often open contact KM2 in parallel with the often open contact of switch SB3. At the same time, the often closed contacts interlocked with the contactor KM1 are disconnected to make the contactor

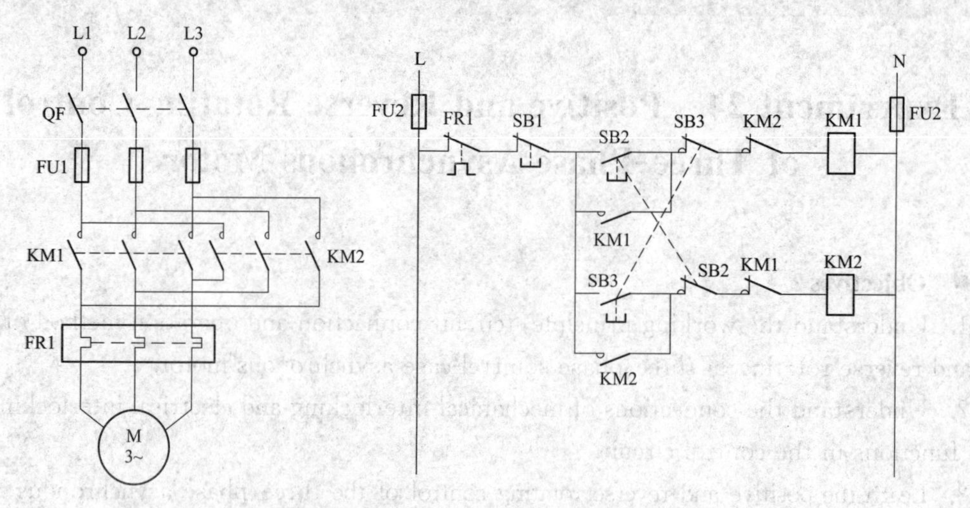

Fihure 24 - 1 The Positive and Reverse Rotating Control Circuit of the Three-Phase Asynchronous Motor

KM1 power off and release. The motor rotates reversely. To stop the motor from running, press the switch SB1.

● Equipment

Equipment is shown in Table 24 - 1.

Table 24 - 1 Equipment

Equipment	Model or Specification	Quantity	Module
AC Power Supply	0～450V Three-Phase 0～250V Single-Phase		QS - DYD3
Three-Phase Squirrel-Cage Asynchronous Motor	220V	1	M14
Relay Contact Control		1	EEL - 57A

● Contents

1. Check the appearance and quality of the experiment equipment.

2. Connect the positive and reverse rotating control circuit of the three-phase asynchronous motor with double interlocking control of buttons and contactors according to Figure 24 - 1. Connect the main loop first, then the control loop. After self-check and the teacher's approval, close the switch and carry out the experiment.

3. Carry out *Positive-Reverse-Stop* operation.

(1) Adjust the thermal relay value to 1. 0A.

(2) Close the leakage protection circuit breaker and air switch QF to add three-phase power supply.

(3) Press the button SB2 and observe the working situations of the motor and each contactor.

(4) Press the button SB3 and observe the motor working situation.

(5) Press stop button SB1 to cut off the motor control power.

(6) Disconnect the air switch QF to cut off the three-phase power supply

(7) Disconnect the leakage protection circuit breaker to turn off the main power supply.

● **Questions**

1. How do the contactor and the button achieve double interlocking in Figure 24 - 1?
2. What is the advantage of double interlocking compared with single interlocking?
3. Why do we want to achieve double interlocking?
4. In the above experiment, what happens to the motor during the conversion process? What is the difference from the *positive-stop-reverse* process? Analyze the reason.

实验 25　三相笼型异步电动机降压启动的控制

一、实验目的

（1）了解时间继电器的结构，掌握其工作原理及使用方法。

（2）掌握 Y-△启动的工作原理。

（3）熟悉实验线路的故障分析及排除故障的方法。

二、实验原理

电动机正常运行时定子绕组接成△形，而电动机启动时 Y 形接法启动电流小，故采用 Y-△减压启动方法来实现限制启动电流的目的。

启动时，定子绕组首先接成 Y 形，待转速上升到接近额定转速时，将定子绕组的接线由 Y 形接成△形，电动机便进入全压正常运行状态。因为功率在 4kW 以上的三相笼型异步电动机均为△形接法，故都可以采用 Y-△启动方法。

图 25-1 所示为三相异步电动机 Y-△启动自动控制线路。

启动时，合上漏电保护断路器和空气断路器 QF，引入三相电源。按下启动按钮 SB2，接触器 KM1 线圈通电，主触头闭合，且线圈 KM1 通过与开关 SB2 并联的辅助动合触点 KM1 形成自锁，同时接触器 KM3 和时间继电器 KT1 都通电，接触器 KM3 主触点闭合，电动机 Y 形启动。当经过时间继电器设定的一段整定时间以后，时间继电器延时断开动断触点 KT1 断开，接触器 KM3 断电释放，其辅助动断触点 KM3 闭合，同时时间继电器延时断开动断触点 KT1 断开，接触器 KM2 线圈通电，其主触点 KM2 闭合并自锁，且与时间继电器线圈 KT 相连的辅助动断触点 KM2 断开，接触器 KM3 和时间继电器 KT1 线圈断电释放，电动机转为△形运转。如需电动机停止运转，直接按一下按钮 SB1 即可。

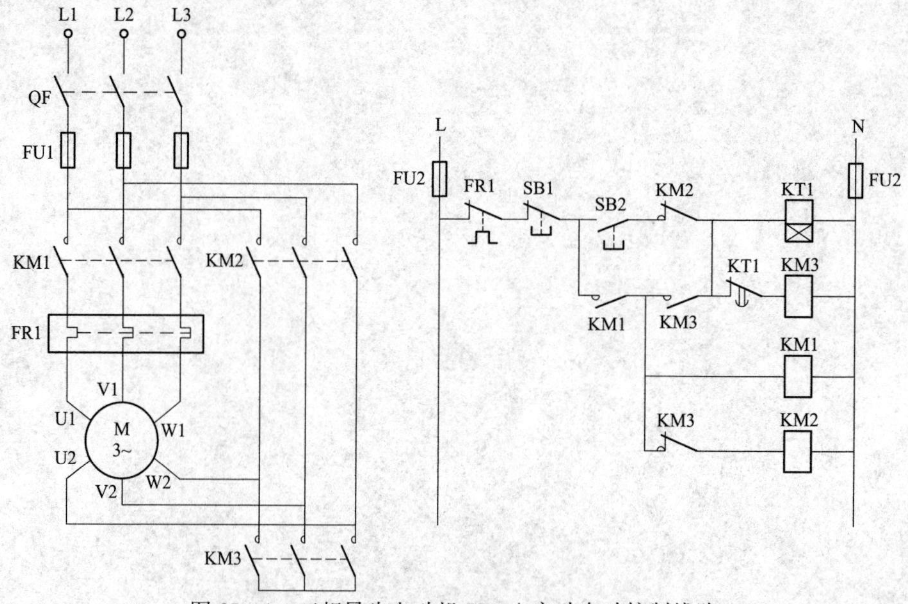

图 25-1　三相异步电动机 Y—△启动自动控制线路

三、实验设备

实验设备见表 25－1。

表 25－1　**实 验 设 备**

设备名称	型号与规格	数量	实验模块
交流电源	0～450V 三相/0～250V 单相		QS－DYD3
三相笼型异步电动机	220V	1	M14
继电接触控制组件		1	EEL－57A

四、实验内容

（1）检查各实验设备外观及质量是否良好。

（2）按图 25－1 三相异步电动机 Y－△降压启动自动控制线路进行正确接线，先接主回路，再接控制回路。自己检查无误并经指导老师检查认可方可合闸实验。

注意：电机运行时间不宜过长。

（3）启动操作。

1）调节时间继电器的延时按钮，使延时时间为 3s。

2）热继电器值调到 1.0A。

3）合上漏电保护断路器和空气断路器 QF，引入三相电源。

4）按下启动按钮 SB2，观察接触器、时间继电器及电动机的工作情况。（注意：电机运行时间不应过长）

5）按下停止按钮 SB1，断开电机控制电源。

6）断开空气断路器 QF，切断三相主电源。

7）断开漏电保护断路器，关断总电源。

五、思考题

（1）分析图 25－1 中电动机是如何实现 Y－△转换的。

（2）在图 25－1 中，如果时间继电器的延时闭合动合触点与延时断开动闭触点接错（互换）、线路工作状态将会怎样？

（3）若在实验中发生故障，分析故障原因。

Experiment 25 Step-Down Startup Control of Three-Phase Squirrel-Cage Asynchronous Motor

● Objectives

1. Understand the structure and the working principle of time relay, and learn how to use it.

2. Learn the working principle of Y-△ starting.

3. Learn the fault analysis and the troubleshooting of experiment circuit.

● Principles

Stator winding is △- connected when the motor is running normally, but the starting current is smaller when the winding is Y-connected. So the Y-△ step-down start is used to limit the starting current.

When starting, the stator windings are Y-connected first when the motor speed is up to close to the rated speed. Change the connection of the stator winding from Y to △, the motor enters the normal running state. Because the three-phase cage asynchronous motors with power above 4 kW are all △-connected, the Y-△ starting method can be used for them all.

TheY-△ startup automatic control circuit of three-phase asynchronous motor is shown in Figure 25-1.

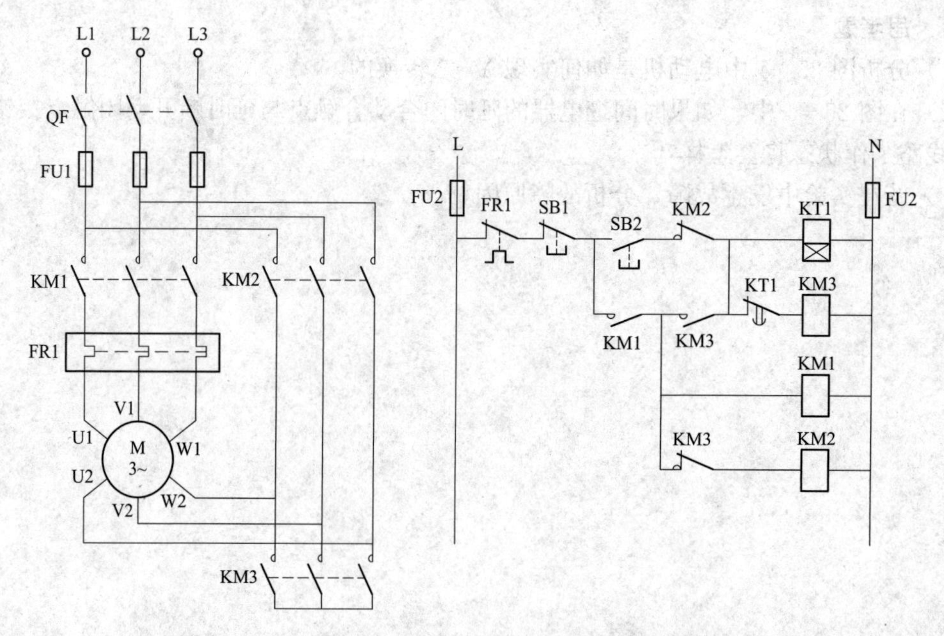

Figure 25-1 The Y-△ Startup Automatic Control Circuit of Three-Phase Asynchronous Motor

When starting, close the leakage protection circuit breaker and the air switch QF to add the three-phase power supply. Press the start button SB2, the coil of the contactor KM1 is electrified, and the main contact is closed. The coil KM1 is self-locked by the auxiliary often open contact KM1 in parallel with the often open contact of switch SB2. At the same time the contactor KM3 and the time relay KT1 are all electrified, and the main contact of the contactor KM3 is closed. The motor starts in Y-connected connection. After a period of setting time of the time relay, the delayed-closing often closed contact KT1 of the time relay is disconnected, and the contactor KM3 powers off and releases and its auxiliary often closed contact KM3 is closed. At the same time the delayed-disconnecting often closed contact KT1 of the time relay is disconnected, and the coil of the contactor KM2 is electrified and its main contact KM2 is closed and self-locked. The auxiliary contact KM2 connected to the time relay coil KT is disconnected, and the contactor KM3 and the coil of the time relay KT1 power off and release. The motor is converted to a △-connected running. To stop the motor from running, press the switch SB1 directly.

● **Equipment**

Equipment is shown in Table 25 - 1.

Table 25 - 1 **Equipment**

Equipment	Model and Specification	Quantity	Module
AC Power Supply	0～450V Three-Phase 0～250V Single-Phase		QS - DYD3
Three-Phase Squirrel-Cage Asynchronous Motor	220V	1	M14
Relay Contact Control		1	EEL - 57A

● **Contents**

1. Check the appearance and quality of the experiment equipment.

2. Connect Y -△ step-down startup automatic control circuit of three-phase asynchronous motor according to Figure 25 - 1. Connect the main loop first, then the control loop. After self-check and the teacher's approval, close the switch and carry out the experiment.

3. Startup operation.

(1) Adjust the time delay button of time relay to set the delay time to 3 seconds.

(2) Adjust the thermal relay value to 1. 0A.

(3) Close the leakage protection circuit breaker and air switch QF to add three-phase power supply.

(4) Press the button SB2 and observe the working situations of the motor, the contactors and the time relay. Note: the running time of the motor should not be too long.

(5) Press stop button SB1 to cut off the motor control power.

(6) Disconnect the air switch QF to cut off the three-phase power supply.

(7) Disconnect the leakage protection circuit breaker to turn off the main power supply.

- **Questions**

1. How does the motor achieve Y -△ conversion in Figure 25 - 1?

2. In Figure 25 - 1, if the delayed-closing often open contact and the delayed-disconnecting often closed contact of the time relay are wrongly connected (exchanged), what is the working state of the circuit?

3. If errors occur in the experiment, analyze the cause of them.

实验 26　三相异步电动机能耗制动

一、实验目的

（1）了解什么是能耗制动。

（2）掌握电动机的能耗制动控制的工作原理、接线方式及操作方法。

二、实验原理

三相异步电动机从切除电源到完全停止旋转，由于惯性的作用，总要经过一段时间。这往往不能适应某些生产机械工艺的要求，如万能铣床、卧式镗床、组合机床等。无论从提高生产效率，还是从安全及准确停位等方面考虑，都要求电动机能迅速停车，要求对电动机进行制动控制。制动方法一般有机械制动和电气制动两大类。机械制动是用机械装置来强迫电动机迅速停车；电气制动实质上是在电动机停车时，产生一个与原来旋转方向相反的制动转矩，迫使电动机转速迅速下降。

能耗制动是在电动机脱离三相交流电源之后，定子绕组上加一个直流电压，即通入直流电流，利用转子感应电流与静止磁场的作用以达到制动的目的。根据能耗制动时间控制原则，可用时间继电器进行控制；也可根据能耗制动原则，用速度继电器进行控制。

图 26－1 所示为能耗制动的控制线路。

启动时，合上漏电保护断路器和合上空气断路器 QF，引入三相电源。按下启动按钮 SB3，接触器 KM1 的线圈通电，主触头 KM1 闭合且线圈 KM1 通过与开关 SB3 并联的辅助动合触点 KM1 实现自锁，并和接触器 KM2 形成互锁，电动机开始运转。当按下按钮 SB2 后，接触器 KM2 的线圈通电，其主触头闭合且线圈 KM2 通过与开关 SB2 的动合触点并联的辅助触点 KM2 实现自锁，同时其对接触器 KM1 的互锁动断触点 KM2 断开，使接触器 KM1 断电释放，电动机进入能耗制动状态，同时时间继电器 KT1 线圈通电。当经过时间继电器一段延时后，其动断触点断开，接触器 KM2 线圈断电释放，能耗制动结束。

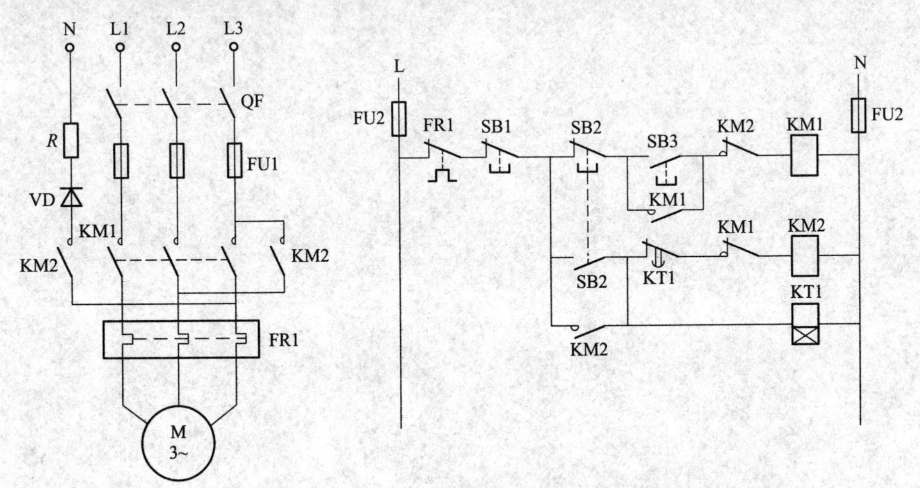

图 26－1　异步电动机能耗制动的控制线路

三、实验设备

实验设备见表 26－1。

表 26－1　实　验　设　备

设备名称	型号与规格	数量	实验模块
交流电源	0～450V 三相/0～250V 单相		QS－DYD3
三相笼型异步电动机	220V	1	M14
继电接触控制组件		1	EEL－57A

四、实验内容

(1) 检查各实验设备外观及质量是否良好。

(2) 按图 26－1 三相笼型异步电动机能耗制动控制线路进行正确接线，先接主回路，再接控制回路。自己检查无误并经指导老师检查认可方可合闸实验。

1) 热继电器值调到 1.0A，调节时间继电器的时间参数，将其设置为 1s。

2) 合上漏电断路器及空气断路器 QF，引入三相电源。

3) 按下启动按钮 SB3，观察电动机、时间继电器及各接触器的工作情况。

4) 按下按钮 SB2，观察电动机、时间继电器及各接触器的工作情况。

5) 断开空气断路器 QF，切断三相主电源。

6) 断开漏电保护断路器，关断总电源。

五、思考题

(1) 分析图 26－1 中是如何实现能耗制动的?

(2) 在上述实验中，电阻 R 的大小以及时间继电器的延时长短对制动效果会产生什么影响?

(3) 比较不同制动方法的平均制动时间及制动效果，简述不同制动方法的使用场合。

(4) 若实验中发生故障，画出故障线路，分析故障原因。

Experiment 26　Energy Consumption Braking of Three-Phase Asynchronous Motor

- **Objectives**

1. Understand the energy consumption braking.

2. Learn the working principle, the connection and the operation of energy consumption braking control of the motor.

- **Principles**

For a three-phase asynchronous motor, it takes a period of time from the cutting off of the power to the complete stop of rotation due to the effect of inertia. This characteristic is often unable to meet the requirements of certain production machinery, such as universal milling machine, horizontal boring machine, combined machine tool, etc. Whether to improve production efficiency or safety or stopping to an accurate position, the motor is required to stop quickly and have braking control. There are two major types of braking methods: mechanical braking and electrical braking. Mechanical braking is using a mechanical device to force the motor to stop quickly and electrical braking is to produce a braking torque that is opposite to the original rotation direction at the time the motor is stopped to force the motor to speed down quickly.

Energy consumption braking is adding a DC voltage to the stator winding after the motor is out of three-phase AC power supply. In other words, let the DC current flow into the stator circuit and use the interaction between rotor induction current and the static magnetic field to achieve the purpose of braking. According to the principle of energy consumption braking time control, a time relay can be used to control energy consumption braking or according to the principle of energy consumption braking, a speed relay can be used to control energy consumption braking.

The control circuit of energy consumption braking is shown in Figure 26 - 1.

When starting, close the leakage protection circuit breaker and the air switch QF to add the three-phase power supply. Press the start button SB3, the coil of the contactor KM1 is electrified, and the main contact is closed. The coil KM1 is self-locked by the auxiliary often open contact KM1 in parallel with the often open contact of switch SB3 and form a double interlock with the contactor KM2. The motor starts running. When the button SB2 is pressed, the coil of the contactor KM2 is electrified, and the coil KM2 is self-locked by the auxiliary often open contact KM2 in parallel with the often open contact of switch SB2. At the same time, the often closed contact KM2 interlocked with the contactor KM1 is disconnected, and the contactor KM1 powers off and releases. The motor enters the energy consumption braking state, and the coil of the contactor KT1 is electrified. After a period of

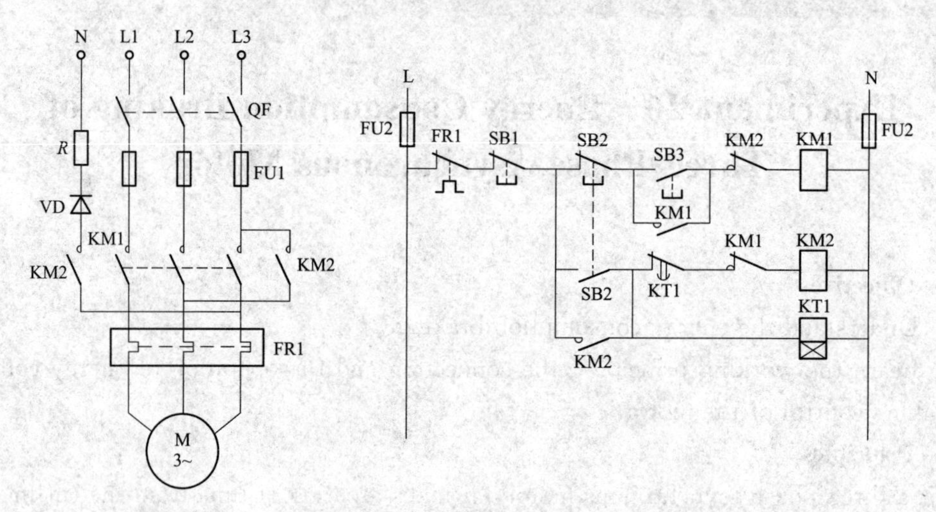

Figure 26 - 1 The Control Circuit of Energy Consumption Braking of Three-Phase Asynchronous Motor

time delay of the time relay, its often closed contact is disconnected, and the coil of the contactor KM2 powers off and releases. The energy consumption braking is finished.

- **Equipment**

Equipment is shown in Table 26 - 1.

Table 26 - 1 Equipment

Equipment	Model or Specification	Quantity	Module
AC Power Supply	0～450V Three-Phase 0～250V Single-Phase		QS - DYD3
Three-Phase Squirrel-Cage Asynchronous Motor	220 V	1	M14
Relay Contact Control		1	EEL - 57A

- **Contents**

1. Check the appearance and quality of the experiment equipment.

2. Connect the control circuit of energy consumption braking of three-phase asynchronous motor according to Figure 26 - 1. Connect the main loop first, then the control loop. After self-check and the teacher's approval, close the switch and carry out the experiment.

(1) Adjust the thermal relay value to 1. 0A. Adjust the time delay button of time relay to set the delay time to 1 second.

(2) Close the leakage protection circuit breaker and air switch QF to add three-phase power supply.

(3) Press the start button SB3 and observe the working situations of the motor, the contactors and the time relay.

(4) Press the button SB2 and observe the working situations of the motor, the contactors and the time relay.

(5) Disconnect the air switch QF to cut off the three-phase power supply.

(6) Disconnect the leakage protection circuit breaker to turn off the main power supply.

- **Questions**

1. How is energy consumption braking achieved in Figure 26 - 1?

2. In the above experiment, what influence do the resistance of the resistor R and the delay length of the time relay have on the braking effect?

3. Compare the average braking time and braking effect of different braking methods and briefly describe the applications of different braking methods.

4. If errors occur in the experiment, draw the circuits of the errors and analyze the causes.

参 考 文 献
Bibliography

[1] 邱关源. 电路. 5版. 北京：高等教育出版社，2006.
[2] 颜湘武. 电工测量基础与电路教程. 北京：中国电力出版社，2011.
[3] 宇秀贞. 新编电工实验指导书. 北京：中国电力出版社，1999.
[4] Davie E. LaLond，John A. Ross，Experiments In Principles Of Electronic Devices And Circuits. New York：Delmar Publishers Inc.，1994.
[5] James W. Nilsson，Susan A. Riedel，Electric Circuits. 9th ed. Upper Saddle River，New Jersey：Pearson Education Inc.，2011.